New
Strategic Factors in
the North Atlantic

Volume No 21 in the Series
Norwegian Foreign Policy Studies

New Strategic Factors in the North Atlantic

Edited by
Christoph Bertram and
Johan Jørgen Holst

under the auspices of
the Norwegian Institute of
International Affairs, Oslo,
and the International Institute
for Strategic Studies, London

Universitetsforlaget
Oslo

IPC Science and Technology Press Ltd.
Guildford

© Universitetsforlaget 1977

ISBN 82-00-05039-4 (Universitetsforlaget)
ISBN 0-902852-69-8 (IPC Science and Technology Press Ltd.)

Cover design: Bjørn Roggenbihl

Published simultaneously in Norway
by Universitetsforlaget, Oslo, as No. 21 in the series
Norwegian Foreign Policy Studies on behalf of the
Norwegian Institute of International Affairs, and
in the United Kingdom by
IPC Science and Technology Press Ltd., Guilford.
on behalf of the International Institute for
Strategic Studies.

The International Institute for Strategic Studies and the Norwegian Institute of
International Affairs are unofficial bodies which promote the scientific study of
international questions and do not express opinions of their own. The opinions expressed
in this publication are the responsibility of the authors.

Printed in Norway by
Joh. Nordahls Trykkeri, Oslo

Preface

The North Atlantic has always been an area of particular security concern for the Atlantic Alliance. Though the area has experienced considerable changes over the past few years, it will see even more in the near future – changes in perceptions, concerns and expectations. The new Law of the Sea will affect aspects of political cohesion and security in a maritime alliance like that of the Atlantic. For the Soviet Union and the United States, as well as for France and Britain, the North Atlantic will, for some time to come, remain an area which lends itself for the deployment of strategic nuclear forces. And the dynamics of modern maritime technology will be most visible in their military and security consequences in this area.

This book examines the range of new and traditional factors of strategic significance in the North Atlantic. It was conceived at a Conference held in Iceland in October 1975, jointly organized by the International Institute for Strategic Studies and the Norwegian Institute for International Affairs. It was probably the first time that scholars, analysts and policy-makers from all the littoral states of the North Atlantic were able to come together to look at the changing security scene in the area. This was made possible by the financial support of the following institutions: the Icelandic government, the BDM Corporation, the Danish Institute of International Affairs and the Nordic Co-operative Committee for International Politics. Without their help the international effort which this book represents would not have been possible.

London/Oslo   Christoph Bertram
Johan J. Holst

Acknowledgements

Many people made valuable contributions in the form of comments and suggestions. In particular we want to acknowledge the important contributions made by Dr John K. Beling, Admiral Worth Bagley, Mr Màr Elisson, Mr David Greenwood, Mr Arthur P. Hockaday, General Tønne Huitfeldt and Mr Erik Klippenberg. Our sincere thanks go to Ms Ragnhild Lindemann for all her help in the preparation of this book.

Contents

CHRISTOPH BERTRAM

New Strategic Factors in the North Atlantic

Major maritime developments have meant that some areas have acquired special significance, and of these the North Atlantic is the ocean region with the most direct political and strategic implication for the Western (North Atlantic) Alliance. It is, therefore, unwise to claim that there are particularly new developments in this region. After all, there are very few sudden and dramatic changes in the world that still seem dramatic and sudden in retrospect. None of the outstanding current developments in the North Atlantic to be discussed in this chapter are, in themselves, entirely new: the build-up of Soviet naval forces; the impact of modern military technologies; and the implications of the heightened attention to ocean resources and the emerging Law of the Sea. But what makes them new, and major strategic factors, is the speed of change in each of these areas, and the combined effects of this change. The consequences can be profound both for the sea area of the North Atlantic, and for the military and political security of Europe.

THE SOVIET NAVY

The Soviet Navy has been a major military force for some time, as is only natural for a country with a long coastline and a problem of communication and transport. What is new, however, is that the navy of the Soviet Union has ceased to be a primarily coastal force and has become a force of power projection; moreover, in the North Atlantic, Soviet deployment patterns, after years of ambiguity, seem to have reached a certain consolidated level which offers some indications of their military and political purpose.

First, the Soviet Union continues to assign the majority of her combat forces to the two Northern fleets. This is most marked in the number and types of submarines, particularly in the Northern Fleet. As Robert Weinland points out in this book, the number of submarines has declined in all Soviet fleets, with the exception of the Northern one.

9

This is largely due to two factors: the introduction of modern ballistic missile-launching submarines, the majority of which are with the Northern Fleet; and the fact that Soviet surface vessels in the area remain highly vulnerable to NATO countermeasures, while it is much more difficult to deny submarines access to the Atlantic. The emphasis on the Northern Fleets in overall Soviet deployment is again the result of two considerations – the primary importance the Soviet Union attaches to the European theatre, and to the circumstances of geo-strategy – if the Soviet Union had firm control over other ports with easier access to the launch points for strategic-missile submarines, the emphasis on submarine-launched ballistic missiles (SLBM) of the Northern Fleet would probably decline.

Second, Soviet naval activity in the North Atlantic seems to have found a certain rhythm and level, after the initial build-up period. But what can be deduced from this for the future? There is a small core force of steady deployment in the Atlantic, on station more or less continuously and increased on specific occasions. Yet it is surprisingly small if compared to the total number of vessels theoretically available. There may be two explanations for this. One would be that the Soviet Union cannot see much advantage in maintaining a significantly higher level of forces on steady deployment in peacetime, since from a military point of view massive deployment in peacetime is unnecessary and costly, and from a political point of view demonstrations of force are most effective in a crisis (and the smaller the routine deployment the more powerful the political impact of deploying additional vessels). The other explanation would be that the Soviet Union may not at this stage have at its disposal a sufficiently seaworthy navy for large-scale continuous deployment, and that the core forces represent the elite of the navy but are neither representative of its overall quality and performance, nor an indication for future deployment. The second explanation is very plausible – and since Soviet naval vessels have not yet been tested in combat there is little else to go by than plausibility. Building up a naval force is not merely a matter of accumulating hardware but the much more demanding and time-consuming task of creating a corps which displays co-operation, similar instincts, reliability on judgement in an emergency – in this sense, tradition. Many of the more fashionable assessments of the Soviet Navy seem to forget this, just as they seem to disregard the discrepancy inherent in a country of the notorious inefficiency of Soviet Russia – allegedly producing nothing but the most efficient submarines and surface vessels. If the level of Soviet naval deployment is not significantly higher at the moment, this may well be the result of these shortcomings rather than of design, and the Russians

10

may be doing no more than making a virtue of necessity. In this case, it would be premature to assume that the present level of deployment will also be the future level. What we are seeing today may – from the Soviet perspective – be not an optimum but a minimum deployment.

The third aspect of current Soviet naval activities in the North Atlantic is that it has not generally been used to influence political events in the area. There are, it is true, two exceptions to the rule which Weinland discusses in greater detail. But on the whole, and comparing the Soviet image in the West with actual Soviet performance, the Soviet Navy in the North Atlantic has tended to display not a demonstration of military threat and political influence but rather what Americans would call a 'low profile'. If, in spite of this, Soviet naval activity in the area has received a good deal of anxious Western attention, this has more often been the result of exaggeration by the observer than of imposition by the actor.

The skilful and effective use of sea power for political influence is one of the most difficult tasks, despite the general enthusiasm naval planners and politicians often seem to have for it. We all know that the silhouettes of warships over the horizon have some kind of effect on the coastal observer; but given the complexity of human reactions, and of the reactions of human society, it is very difficult to say in advance what exactly the effect will be: resigned acceptance, indifference or heightened opposition. The display of military force is an ambiguous instrument of influence – not because it is *without* influence, but because it is so difficult to predict the result the influence will have, once applied.

Nevertheless, the Soviet Union's past reluctance and lack of skill in using a show of naval force to influence political decisions of the coastal states should not be a cause for complacency. For one thing, the impact of a show of force on political behaviour is most direct in a situation of political crisis – suppose a succession crisis in Yugoslavia when the Soviet Union might wish to demonstrate to the Northern members of the Alliance the disadvantage of opposing Soviet action. The second point is that the low profile of continuous Soviet deployment leaves a great deal of reserve forces to be displayed in order to make a political point. Third – and perhaps most important – the Soviet Union is aware both of the fact of her geopolitical disadvantages and the difficulty of overcoming them by military means. It makes much more sense for her to try to influence the behaviour of countries like Denmark or Norway by political pressure before a military contingency arises. Though there is the risk that the effect of naval force demonstration may backfire, the prize of successful pressure is very high indeed, as are Soviet reserves for the show of force that might produce this pressure. It is for these reasons that past and present performance by Soviet naval forces in the region offer little firm ground on which to base a projection of the future.

NEW NAVAL TECHNOLOGIES

The second new factor of strategic importance is the range of modern naval technologies currently being introduced into the arsenals of major navies. While it is impossible to go into the whole detailed catalogue of new naval weaponry, some major trends are becoming apparent.[1]

First, the vulnerability of surface ships to strikes from the sea, the air or from land has increased dramatically. Because surface vessels are visible, and satellite and other types of reconnaissance are capable of detecting them over wide stretches of water, hiding becomes more difficult, and precision-guidance enables weapons to reach targets that can be defined. Precision is no longer a matter of the distance from land: the time has passed when distance alone meant inaccuracy and therefore protection. Modern delivery systems, land-based, sea-borne or air-borne, have acquired a very high degree of accuracy and this will be improved further as some of the existing shortcomings are ironed out, such as weather dependence or over-the-horizon guidance. Navies that want to stay afloat and provide sea control or project power beyond their coastal zone will try to meet this development in two ways: a greater emphasis on submarines and aircraft for sea control, and – since these are highly limited in their relevance for power projection – emphasis on self-defending units and task forces. This is not necessarily an argument against the big ships; the requirement for aircraft carriers, for instance, will depend on the mission to be performed and on the money available. But it does argue against concentrating the naval effort on big ships alone.

Secondly, the ability to strike at conventional military targets on land from sea-based platforms has also increased considerably. The most obvious example of this is the modern cruise missile: launched from a torpedo tube, it can reach targets over a distance of up to 2,000 nm. with a very high degree of accuracy. As long-range weapons systems installed in fixed positions on land become more vulnerable to pre-emptive strikes as a consequence of improved accuracies, seamobile delivery systems with the ability to strike at major targets on land become a more interesting strategic option for the larger navies. For the navies of the two superpowers at least, this will increasingly blur the – admittedly somewhat artifical – distinction between tactical and strategic use of naval power as the theatre for which

12

naval weapons deployment can be relevant will extend far beyond the horizon.

Thirdly, modern naval technology favours the practice of sea denial by coastal states – that naval mission which consists of preventing other naval forces from imposing their control in waters adjacent to a littoral state. With the development of relatively cheap and highly accurate anti-shipping missiles and what appears a striking improvement in mine technology, the ability of even major navies to roam with impunity in the coastal areas of a hostile state will be significantly circumscribed. What is more, and this is perhaps the most striking consequence, effective sea denial can be performed by small states against the navies of big states: Kuwait, or Bangladesh, or Norway or Denmark could muster a very formidable deterrent against Soviet or American naval forces operating in adjacent waters. This will not be sufficient for the policing of the new 200-mile economic zones which the Law of the Sea Conference is likely to bestow on coastal states, but it will represent a major curb on the ability of other naval forces to project power into a nation's own maritime backyard.

The overall picture which emerges from these trends for the North Atlantic area is ambiguous. The projection of hostile power through naval forces will become more circumscribed; Soviet naval pressure against, say, Norway becomes more deterable in theory, but in practice this will depend decisively on the will of a small country to deter. At the same time, the projection of benevolent naval power will also become more difficult; the vulnerability of major surface vessels means that, say, an American task force in the North Atlantic, intended to reassure other NATO countries in the North of America's support, will provide a less impressive demonstration of strength than in the past. Similarly ambiguous is the effect of the greater degree of self-reliance of coastal states in the defence of the adjacent sea areas: theoretically, this implies a decreasing need for Alliance support and greater self-sufficiency in defence; in practice, however, this task can only be performed if access to modern technologies is assured, and this is most promising in an alliance which includes the United States. Finally, while the improvement of coastal defence offers a real chance to deny coastal areas to hostile fleets, the improvement of missile range and accuracy means that an enemy wanting to destroy targets on shore does not need to operate in the coastal areas but can attack from afar, with a greater chance of protection and an equal chance of hitting the land target.

RESOURCES AND THE LAW OF THE SEA

The third major new strategic factor in the North Atlantic is the growing attention given to ocean resources and the consequences for international law this entails. The North Atlantic has been a familiar resource for fish; it has become, in addition, a resource area for oil and gas.

This is perhaps less important in itself than in its political and psychological impact on the littoral states. The new fascination with the riches of the North makes the inhospitable Norwegian Sea, the cold, wet and wind of the North Atlantic, the barren stretches of ice and snow of Greenland, the Arctic or Svalbard often appear like the new frontier of popular imagination and political expectations – particularly after the energy crisis.

The ocean resources in the North Atlantic are indeed important; they will affect the economic performance, social structure, domestic politics and international standing of those lucky states with a lot of coastline. Many of the countries in the region have not suffered from a shortage of energy resources in the past, yet they tend to see the answer to many of their problems in the additional resources of the North Sea. The psychological impact of the new riches under the sea may, therefore, be by far more important and politically consequential, as Finn Sollie reminds us in this book. The Arctic rush is on, and it is bound to affect political expectations and sensitivities.

The fascination of the new resources will be displayed and reinforced by the new Law of the Sea. The introduction of a general Exclusive Economic Zone of 200 miles for littoral states will give additional proprietal respectability to the exploitation of ocean resources. It will serve to reduce the area of unrestricted sea space for navies and fishing boats alike, and what is intended as an area of functional exploitation may become, in time and in the view of governments, an area of sovereign control, or at least an area to keep out hostile navies by law if not by force. The territorialization of the seas is underway, however unsuited the fluid media of the ocean is for the precise demarcations of land.

The strategic implications are twofold. First, the room for manoeuvre of navies will be increasingly circumscribed; second, there will be new causes for conflict.

On the face of it, the new Law of the Sea will distinguish between territorial waters of 12 nm, where traditional restrictions for naval vessels of other countries apply, and exclusive economic zones of 200 nm which are, in terms of the old law, the high seas. But, as David O'Connell remarks in this book, the high seas will be much less free: littoral states will insist on pollution control; they will want to monitor scientific research (which will restrict use for military research); and they will want to have some capability to police their new sea terrain. Surface vessels will, at least, have to conform to the rules proclaimed by the littoral state. The new laws will affect the old navies – perhaps not by producing effective naval arms control, but effective naval arms hindrance since the space of blue water is shrinking.[2] Even the designation of the high seas area (beyond 200 miles) as the common heritage of mankind contains, as Professor O'Connell points out, the germ of progressive neutralization of the oceans.

In the North Atlantic this will above all affect the Soviet Navy. Maritime alliances may take on a new significance when sovereign states grant the use of their sea space to their friends (as they tend to do with their air space), and this gives a particular advantage to the American Navy in the North Atlantic. Just as lack of air support remains one of the most serious handicaps for the Soviet Navy, so the lack of maritime allies will mean, under the rules of the emerging new law, a handicap for naval operations in peacetime.

This is bound to generate counter-pressures and, in the North Atlantic, counter-pressures from the country whose naval forces will be most affected: the Soviet Union. It may well be that one of the motives for present Soviet deployment in the North Atlantic and the Baltic is an attempt to create a pattern of precedents, which littoral states will have to challenge once the new law is agreed on, but which they may lack the power and the will to do. International law is not self-executing; not self-enforcing. It needs instruments of force to back it up. This is a costly business, and one which not many coastal states will wish to incur. Moreover, a nation's lack of will and of means to defend her zone of exclusive economic exploitation at sea may be seen by a more powerful navy as a perfect justification for projecting power – to fail to demonstrate sovereignty could be interpreted as resignation to superior military power. This is one of the almost unsolvable dilemmas for Norway in her dispute with the Soviet Union over the demarcation of the continental shelf in the Barents Sea, on which they hold different positions. But if there is no agreement will Norway stand by her legal position and start drilling for oil in ocean areas the mighty Soviet neighbour claims for herself? And if Norway does not, will she then not recognise Soviet claims implicitly?

This exemplifies another possible impact of the new Law of the Sea and the emphasis on resources: that of providing new sources for conflict. This is partly the fault of the law itself which, to go by the draft text for the Law of the Sea, refrains from offering any precise guide for the demarcation of adjacent sea zones and so gives – as always when there is a wide margin for interpretation – a bonus to the pressure of the superior military power. But whatever the law, the sea itself is not a medium which lends itself to the clear demarcations and definitions of national boundaries of the land. Moreover, the new Law of the Sea provides for an extension of national jurisdiction over areas that until now were unrestricted. It has not been a vacuum to be simply appropriated by the nation-states; the extension of national jurisdiction means that those activities which have hitherto been unchecked will now be subjected to national rules. There will inevitably be disputes about demarcation (what belongs to whom), about the extent to which national jurisdiction does apply (how much can the littoral state claim?), and to what extent traditional rights must be respected, even under the new legal regime. There will be problems of policing the new areas against intruders and against violations of the rules proclaimed.

This will inevitably be a fluid, fluctuating, untidy situation, and navies will therefore be wise to stay away from the task of policing 200-mile zones if they possibly can. But will they? Can the states of Europe afford to spend money both on navies and on coast guards which operate up to 200 miles off the shore? There will be a temptation here for admirals as well as politicians. Admirals may be tempted to profit from the public interest in ocean resources in order to obtain generous budgets in the hope that they can maintain navies with only a secondary policing role. Politicians, on the other hand, will wonder what navies are for if not for the protection of coastal resources – after all, for much of 1975 and 1976, British frigates were protecting British trawlers around Iceland instead of shadowing Soviet manoeuvres. The result could well be that only the superpowers will be able to retain anything like a blue water navy, while for the rest the task of coastal patrol may become the dominant one in naval force allocation and planning – a trend further strengthened by the apparent advantages offered by new naval technologies for coastal defence.

But would this be a desirable development? The primary function of naval forces must be to protect the security of a country, and national security is not identical with the security of oil rigs and other offshore installations, nor with the maintenance of a credible claim to semi-sovereignty over wide stretches of ocean. By gaining *new* responsibilities and rights, littoral states are by no means losing their *traditional* security problems. The need to secure supplies and military reinforcements in a European land-war cannot be met by a flotilla of coastal patrol boats, and the determined effort by a hostile power to establish a bridgehead ashore by means of amphibious landings cannot be jeopardized by forces designed to keep foreign fishing fleets

16

from violating the national exclusive economic zone. Throughout the fascination with the new frontier of ocean resources and the expansion of littoral jurisdiction, a very careful and sober analysis will be required to retain the priority for national security and to assure that efforts to deal with the new problem of protecting ocean resources do not conflict with the familiar one of protecting the state and the land.

CONSEQUENCES FOR EUROPEAN SECURITY

What are the consequences that follow from these three major new factors for European security? After all, if it were not for the security of Europe and the operation of the sea-based nuclear deterrent forces, there would not be much strategic significance to the North Atlantic. As the range of strategic submarine-launched missiles increases, the area becomes less important as a staging theatre for SLBMs, especially to the U.S. The impact on European security of developments in the area will, therefore, become the primary one. For obvious reasons it is the impact on the chances and likely outcomes of conflict on the land-mass of Europe that is decisive; war in Europe is unlikely to be started at sea, and military considerations on land will determine the use of naval forces at sea. Therefore the actual balance of forces in Europe, and the balance of reserves and reinforcements from outside the war theatre, will to a great extent define the relevance of naval activity in the Atlantic.

The European balance between East and West today is not, at the moment, one of unambiguous Soviet and Warsaw Pact superiority, but of a superiority in the number of men under arms, particularly in combat units, and there is also a very marked numerical superiority in tanks and aircraft. But this is still, to some extent, offset by NATO advantages – in aircraft, anti-tank strength, better technology.[3] What is worrying today is not the actual state of the balance but the future consequences of the apparently uninhibited momentum of Soviet arms production, which could gradually nibble away the advantages NATO defence possesses at the moment. It could also strengthen the traditional tendency in Soviet thinking – that military might produces sizeable political dividends.

On the whole, the real impact of further increases in Soviet military strength is indeed likely to be political rather than military. The risk of nuclear escalation of any conflict in Europe remains a powerful deterrent to military aggression and will continue to be so as long as both superpowers are present in Europe and are eager to avoid direct military conflict. The political rather than the military effect also applies in the North Atlantic. The degree to which, for instance, the enlarged Soviet Navy can really intercept military reinforcements in the event of a major European crisis or war will remain relatively low for some time, given the vulnerability of Soviet surface vessels and the overriding priority of their naval forces to protect their strategic submarines. But if Europeans and Americans were to begin to doubt whether American reinforcements would get through, the political effect would be to seriously weaken the cohesion, and with it the credibility, of the Alliance.

The Alliance retains decisive advantages in the North Atlantic. The Soviet Navy has to rely on strategically vulnerable and geographically disadvantageous ports for its ships. NATO can stage effective air dominance using its bases from Norway to Iceland, Britain and North America. Additionally, the new naval technologies tend to favour an alliance of maritime states more than the Warsaw Pact. The new emerging Law of the Sea should in theory inhibit a superpower navy *without* allies in the area much more than a superpower which leads an alliance of essentially maritime countries.

To meet the political challenge to the Alliance and to European security that is likely to arise from these new strategic factors will, however, not be easy. First, the new attention to ocean resources in the North could lead to a neglect of traditional security needs. The 1975/76 Cod War between Britain and Iceland is an obvious and depressing example. As discoveries of oil and gas in the North Atlantic continue, and as the glut in international oil supplies disappears with the recession in the industrialized world, the Cod War, far from being an exception, could become a precedent.

Second, the pull to the North, so strongly felt in many countries in the area, could make it more difficult to maintain what has in the past been the central security focus of the Alliance: the focus on Central Europe. In this respect the otherwise misleading term flanks may apply: as in Southern Europe where the Greek-Turkish dispute, the Middle East crisis and the domestic upheavals deflect security thinking of governments from the Central European theatre, so may the fascinating pull of ocean resources in the North. In future this will make it more difficult to define common security objectives in precise terms for an Alliance which depends, and will continue to depend, on a commonality of security concerns and on the belief that collective security is essential to meet them.

ANDERS C. SJAASTAD and JOHN KRISTEN SKOGAN

The Strategic Environment of the North Atlantic and the Perspectives of the Littoral States

In this survey we intend to take a closer though necessarily brief look at four sets of factors. The first three are basic constituents of the strategic environment in the North Atlantic, and the fourth may come to have an important impact on this environment by changing the perspectives of the littoral communities.

First, however, a note of terminological clarification. The meaning of the term 'North Atlantic' is clearly a matter of convention, but here we shall take the liberty of departing somewhat from the most common usage by referring only to the northern part of the area between Europe and North America when using the term. In our usage, moreover, the North Atlantic will also include the Norwegian Sea.

The first set of factors that we are going to look at concerns the physical environment: those features of geography and climate which could either constrain or compel military action in the North Atlantic region. The second set consists of the indigenous political conditions and the prevailing perspectives of the various littoral communities which will obviously have a direct bearing on operations involving the use of local territory in the area. At present these conditions favour NATO far more than the Soviet Union, but, unlike features of the physical environment, they are susceptible to the influence of other powers and liable to change. The third set of factors is constituted by the nature and scope of outside military interests and engagement, especially naval engagement in the area. Here we will be paying particular attention to the interests and engagement of the two superpowers, both of whom border the North Atlantic on opposite sides. Finally we shall have a brief look at problems and prospects concerning the exploitation of natural resources in the area, both protein and hydrocarbon. Though not obviously a part of the strategic environment, the politics of the resources in the North Atlantic region present problems which could make profound inroads on the politico-military picture, imposing strains on traditional relations and offering new opportunities for influence from the outside.

THE PHYSICAL ENVIRONMENT

The North Atlantic is bordered by the British Isles, Norway and the Barents Sea to the east and by Newfoundland, Labrador and Baffin Island to the west. Baffin Bay east of the Canadian archipelago and the Greenland Sea west of Svalbard form its two northernmost extensions, between which the vast Greenland land mass extends further south to the latitude of 60°N. Between the Norwegian Sea in the north-east and the rest of the North Atlantic lie Iceland and the Faroe Islands which, with Greenland across the Denmark Strait, form a link of islands between the North American and European continents. This is the Greenland–Iceland–United Kingdom (GIUK) gap, which is important in several respects for naval operations in the North Atlantic.

The waters to the south of Greenland and Iceland and in the central parts of the Norwegian Sea are deep – mostly over 2,000 metres. Along most of the coasts, however, the continental shelf is rather wide and the waters relatively shallow. The waters of the North Sea and the Barents Sea are also shallow, unlike the very deep Polar Basin to the north. From Greenland, an underwater ridge stretches via Iceland and the Faroes to Scotland.

Along the coasts of Baffin Island and Labrador and along the eastern coast of Greenland currents carry a lot of drifting ice and cold water with a low salinity southwards from the polar regions. As a result, most of Baffin Bay and parts of the Labrador coast, as well as the western part of the Norwegian Sea, are covered by pack ice during the winter and spring, leaving the entire east coast of Greenland inaccessible to surface ships during that time of the year. The southern and eastern parts of the North Atlantic, on the other hand, are strongly influenced by the Gulf Stream, which divides west of the British Isles on its way north-eastwards. One part of it proceeds towards the water south of Iceland – dividing again into two new branches, one of which flows north through the Denmark Strait round almost all of Iceland, where it maintains a comparatively mild climate. The other branch continues submerged below the East Greenland current, to the southern tip of Greenland, where it surfaces, keeping the southern part of the west coast of Greenland open to surface ships all year round; the climate here is also markedly milder than the extremely rough climate of east and north Greenland and of the Canadian archipelago. West of the British Isles the rest of the Gulf Stream proceeds northwards, causing a surprisingly mild climate at high latitudes along the northeastern part of the North Atlantic. This branch of the Gulf Stream also divides into two off northern Norway, with one stream continuing northwards along the edge of the continental shelf west of Bear Island and Spitsbergen, keeping the waters there open all year. The other branch bends eastwards, keeping the Barents Sea south of Bear Island – and usually as far east as the mouth of the White Sea – free of pack ice, even in the spring, when the conditions are most severe. In the west, the line of maximum extension of pack ice normally goes from the north coast of Spitsbergen past the Norwegian island of Jan Mayen before bending southwest and continuing through the Denmark Strait, well off the east coast of Greenland to its southern tip, Kap Farvel. During summer and early autumn the pack ice retreats from practically all the North Atlantic.

Waters covered by pack ice are not necessarily unnavigable for surface vessels, provided ice concentration is not too high, but it makes sailing slow, as well as risky, for ships not specially constructed for travelling in pack ice waters. If severe ice coats the superstructure of the ship, it can impair stability, and a sudden increase in the concentration of ice prevents the ship sailing further in any direction. Nuclear-powered submarines may find pack ice waters suitable for concealment for precisely the same reasons that make them unfavourable for surface ships. The ice not only provides protection against surface vessels, but partly against anti-submarine (ASW) aircraft too, leaving the submarine mainly exposed only to underwater ASW systems – either underwater detection devices or enemy hunter-killers. For Soviet submarines, pack ice may even render modern American hunter-killers less formidable, since the Americans' most deadly weapon, the *Subroc* torpedo missile, is designed to travel part of the distance to its target through the air, thus having to break the surface of the water twice.

As to temperature and salinity, the water is relatively homogeneous in the eastern parts of the North Atlantic, but further to the west in the waters east and south of Greenland the cold polar water mixes with that of the Gulf Stream, which results in variations of temperature and salinity, and submarines can take advantage of these to avoid detection. In the Barents Sea there is also much more mixing of cold and temperate water than there is further south-west.

All the territories adjacent to the North Atlantic, except the British Isles, are thinly populated. The importance of fishing and related industries is a common feature of each of the littoral communities. On the eastern coast of Labrador the population totals less than 20,000; fishing and some mining constitute the two most important industries. Greenland has a population of some 50,000 and most of the inhabitants live along the southern part of the west coast, with a few scattered local communities further north and along the southern half of the east coast. In Greenland, too, fishing and mining – though the latter on a small scale – are the main industries. The former accounts for 80 to 90 per cent of export revenues. Since the war hunting, which was the traditional occupation, has declined rapidly, giving rise to considerable unemployment and making the people heavily dependent on large Danish subsidies. The glacier-covered interior is uninhabitable, like the desolate moon landscape in central Iceland. Iceland has a population of 215,000, which is spread all around the coast. The majority of Icelanders live in the south-western corner, and approximately 85,000 – almost 40 per cent – in the capital of Reykjavik. Again, the main industrial activity is based on fishing, which accounts for 70 to 80 per cent of exports. The 40,000 inhabitants of the Faroe Islands, for their part, are dependent on fishing for more than 90 per cent of their export revenues. The Norwegian exports of fish and fish products account for less than 10 per cent of the total, though nearly 40 per cent (1.5 million) of the total Norwegian population lives on the west coast and for them fishing and its related industries are far more important. This is especially true for northern Norway with half a million inhabitants, many of whom are fishermen.

PERSPECTIVES OF THE LITTORAL COMMUNITIES

The perspectives of the communities bordering the North Atlantic are, of course, strongly connected with their domestic policies; those of Britain are fairly well-known, as are those of Labrador, which is a fully integrated part of Canada, so we are going to have a look at Greenland, Iceland, the Faroes and Norway.

All these territories are linked to Europe, constitutionally or physically, but the inhabitants are far removed, both by geography and sentiment, from the political centre of gravity in Europe and for centuries they have been dependent on the oceans, for transportation and resources. In recent years, however, they have become very dependent on overseas imports, especially since the advent of the modern welfare state. Their small populations make it extremely difficult for most of their domestic industries to achieve economies of scale in their production, and the lack of capital prevents them from competing efficiently in the export market. Consequently, in order to finance imports and thus maintain a high standard of living, the littoral communities, with the exception of Norway, have had to rely on their exports of fish and fish products, which makes it of vital importance that they secure a sufficient proportion of the total stock. But, while they have all become well integrated into the international economic system and are affected by its ups and downs, they have nevertheless tried to remain politically uncontaminated by rivalries and conflicts outside the North Atlantic, particularly in continental Europe.

Thus, isolationism and neutralism have been influential factors in domestic politics, although they have been checked in the final formulation of the foreign policy of Iceland and Norway by the experiences in World War II. The urge to remain outside big-power rivalry – and especially to keep its consequences at a distance – has historical sources, but it is due to other factors as well, such as the geographical remoteness of the peripheral communities and, with the exception of Scandinavia, their lack of political affinity with the rest of Europe. Their small size, the constitutional status of some of them and, possibly, the complicated nature of the security problems involved are other factors.

Since the Napoleonic Wars, in which Denmark was on the losing side and had to cede Norway – though not Iceland, Greenland or the Faroes – to Sweden, who was fortunate in belonging to the rival camp, the Scandinavian countries have tried to remain outside Central European conflicts. In effect, and later by explicit formulation, their foreign policies became neutralist, and Norway fell in line with this when she won her independence from Sweden in 1905. Scandinavian neutralism survived the severe test of World War I successfully, giving renewed impetus to the belief in the possibility of keeping the Scandinavian countries, and the Danish dependencies in the North Atlantic as well, outside future big-power conflicts. World War II broke the spell, except for Sweden. Denmark and Norway were occupied by Germany, Iceland and the Faroes by Britain, and Greenland had to ask the United States for military protection, thereby possibly pre-empting a British/Canadian occupation.

Having received a rebuff, the forces of neutralism were not strong enough after the war to prevent Denmark, Norway and Iceland (who had broken away from Denmark during the war) from joining the North Atlantic Treaty as the cold war got worse. But several concessions were made to neutralist sentiments, which partly explains the Danish and Norwegian policy of no foreign bases on their territories in peacetime. (Iceland originally adopted this policy as well, but abandoned it after the outbreak of the Korean War.) Neutralist sentiments are still present in the littoral communities of the North Atlantic region, including the Faroes and Greenland, largely because of the traditional influence of two or possibly three underlying factors. First there is a quite widespread feeling of cultural and political remoteness from continental Europe, demonstrated recently by the Norwegian, Icelandic and Faroese refusals to join the Common Market, and the Greenlanders' opposition to their membership. This attitude is com-

23

mon right across the political spectrum, especially on the Right, where it is often balanced by an interest in efficient defence, requiring co-operation with allies. Secondly, there are the anti-militaristic and anti-imperialistic attitudes – mostly among leftist groupings – which reinforce the interest in non-alignment. Finally, there is the sense of smallness and impotency. They feel they are too small to matter anyway, therefore it is up to others to take care of the problem of security. The best the littoral communities can hope for is not to influence the events but to escape being directly involved; their smallness may seem to improve the prospects of being able to hide away from the turmoils of the world outside. This attitude is particularly strong in the Faroes, who have explicitly asked the Danish government to keep them out of NATO defence arrangements.

The consititutional status of the various littoral communities varies significantly. As independent countries, Norway and Iceland are fully responsible for deciding their own foreign and security policies. The Faroes and Greenland are part of Denmark – though the Faroes have an extensive degree of domestic self-government – and the Danish Government has the exclusive right to formulate the foreign policy of both territories and to look after their security. In some cases this means a foreign policy more tailored to Danish interests and perspectives than to those of the dependencies (at least this is the way some Greenlanders and many Faroese see it), which is a constant source of potential conflict between the Danes and their two dependencies. This is true especially between the central authorities in Copenhagen and the local authorities in the Faroes, where the independence movement has a long tradition and a lot of support, blunted, however, by the continuing need for Danish economic subsidies.

One effect of the uneasy relationship between Copenhagen and the Faroese is that the latter tend to consider foreign policy and security matters mainly within the context of Danish-Faroese relations. Such matters very easily become part of the controversy about the division of authority between the central and local authorities. Thus, the inclusion of the Faroes in NATO, through Danish membership, is considered in the light of the question about boundaries of Danish authority. It is resisted by quite a few Faroese because their local parliament was not consulted before the Danish entry into NATO. The presence of an important radar station not far outside the capital is considered less a matter of defence than one of Danish encroachment, since permission for its construction was not requested from the Faroese parliament, which has repeatedly but unsuccessfully demanded its removal.

Concentration on matters of authority and of Faroese rights in relation to Copenhagen tends to suppress considerations concerning security matters, which are not in any case a Faroese responsibility. The lower the profile maintained by the islands in security policy and the more passive their engagement in foreign policy matters – both Danish responsibilities – the less probability there will be of Danish influence in their affairs. For the present, this is a characteristic feature of the attitudes to security matters in the Faroes, and changes in Greenland in domestic politics as well as in her constitutional relationship to Copenhagen may cause the Greenlanders to regard security matters in a similar way. If that were to happen, the importance of Greenland in the defence of the North American continent and the possibility of increasing Soviet interest in Greenland's territory might cause considerable problems.

A primary objective for all the littoral communities of the North Atlantic is the preservation of their national identity and character. It is sought most vigorously in Iceland and the Faroes, and least actively in Norway, so size may be an important factor; the smaller the community, the greater its efforts seem to be to preserve its special character, and to guard this against destructive influences from outside. If this is so, such efforts will probably be stepped up in Greenland in the future; indeed, there are already some indications that this is happening. The constitutional status may also effect the community's attitude towards preserving its national identity. When national independence has been obtained relatively recently, as in the case of Iceland today, and Norway earlier in the century, sovereignty is probably less easily taken for granted and greater effort is made to guard against trespassing or encroachment or even against certain co-operative efforts with allies, such as the establishment of bases and military installations. The Faroese, who are not independent, would seem to consider it all the more necessary to foster and protect what there is of national identity and integrity. Cultural and social influence from allies is perceived as a more immediate threat than military attack from enemies. Such considerations provided the basic rationale for the Icelandic decision of 1971 to ask the Americans to leave Keflavik.

Except for Norway, none of the littoral communities in the northeast Atlantic has a national armed force, but they are all part of the NATO defence area. And more directly, the military defence of the Faroes is provided by Denmark, while both Denmark and the United States take care of the defence of Greenland. The population of Iceland is too small to be able to sustain a credible, exclusively national defence establishment, and Iceland has refrained from setting up any military forces of her own. The defence of Iceland is ensured in peacetime by American forces at Keflavik on behalf of NATO. The lack of direct responsibility for their own military defence has certain unfortunate effects; with no expertise of their own on military matters, the littoral communities (except Norway) are left with little or no independent ability to evaluate the implications of the increasingly complex security environment. This gives their allies, and especially the bigger ones, a near monopoly of information and know-how and makes it extremely difficult for the small littoral communities to test the wisdom handed down to them. The result may be frustration among the decision-makers and also for the public, who then become apathetic as far as security policy problems are concerned. When questions of national territorial security cannot be avoided, the answers could turn out to be badly improvised and ill-considered.

THE INFLUENCE OF MILITARY INTERESTS IN THE NORTH ATLANTIC

The North Atlantic is vital for the sea lines of communication between North America and Europe. The feasibility of reinforcements from the United States and Canada to Europe, especially to northern Europe, is critically dependent on the use of these lines of communication in an emergency. Accordingly, the Western alliance has strong interests in both sea control and tactical ASW in the area in order to keep the lanes open. Since the advent of the ballistic missile-firing submarine, the North Atlantic has occupied a central position in the nuclear strategies of the two superpowers. For the United States it became an area of deployment as a significant number of American (and later probably British and French) nuclear ballistic-missile submarines (SSBN) were deployed in the North Atlantic. For the Soviet SSBN it became an area of transit en route to and from stations off the eastern coast of North America. Consequently, both sides took great interest in strategic ASW in the area, particularly in its eastern parts. With the increasing range of submarine-launched missiles, the American interest in deployment in the northeast Atlantic and the Soviet need for transit through these waters may gradually decrease. But American interests in strategic ASW in the north-east Atlantic will remain, as will Soviet interests in counter-measures – provided no limitations on strategic ASW are agreed upon. The availability of bases and other military installations on the littoral territories in the North Atlantic makes it relatively easy for the West to attain sea control and to pursue ASW – both tactical and strategic – in the area. The American base at Keflavik, in particular, is of vital importance. The various installations permit the Western allies to carry out extensive and continuous surveillance of the North Atlantic and its points of entry from the east. The Americans and the Norwegians are flying the P–3 *Orion* on maritime patrol missions from Keflavik and Andøya, respectively, while the British operate the *Nimrod* from Scotland. Furthermore, fixed radar installations are scattered through the area. Greenland, Iceland, the Faroes and Norway house *Early Warning* coastal radars and other installations. At Thule air base in northern Greenland, the Americans have constructed a ballistic missile early warning station (BMEWS). On several of the territories there are air bases and field facilities, many of which are equipped to support and replenish advanced fighter aircraft. Thus, the NATO Alliance commands sufficient equipment and installations to exercise sea control in the North Atlantic in peacetime, and the activities and training are naturally geared to denying the Soviet Union the capability of projecting sea power into the North Atlantic in times of military conflict. Some of the air bases, with Keflavik as the prime example, would also be of great significance as staging posts for airborne reinforcements to the northern flank.

In recent years the build-up of the Soviet Navy has presumably given rise to a new interest in the north-east Atlantic on the part of the Soviet Union. The build-up has manifested itself in the north more than anywhere else, and the basic reason for this would seem to be the favourable geographical location of the home ports for the Northern Fleet on the Kola peninsula. From both the Baltic and the Black Sea, Soviet naval vessels have to pass through narrow straits, not controlled by the Soviet Union or her allies, in order to reach the high seas of the western hemisphere. But from the ports on the Kola peninsula passage is virtually free through the Norwegian Sea into the Atlantic Ocean. This has been essential for the SSBN, but for surface vessels as well access to the high seas is of vital importance to a rising naval power. It appears likely, therefore, that the Soviet Union will attempt to establish sea control in the Norwegian Sea in times of crisis. The objective collides head-on with the Western interest in securing the sea lines of communication across the North Atlantic. The conflict emphasizes the importance of the military installations on the littoral territories of the area and that of the GIUK gap as a barrier to the Soviet Northern Fleet, but the allied military installations on the littoral territories of the North Atlantic and related activities are still subject to opposition and resentment from some popular groups, as is NATO membership – either directly or through Danish membership, and such opposition may find increasing support during a period of East-West detente, although there are no signs yet to that effect. But the present configuration of interests relating to the North

Atlantic and the strategic significance of the area make a policy of aloofness a less viable solution for the littoral communities than ever. None of them, however small in population or territory, can expect to escape from the calculations and aspirations of outside powers. The problem presenting itself to all members of NATO – not least to the littoral communities of the North Atlantic – is how to discourage Soviet aspirations in the area. None of the littoral communities can solve this problem successfully by adopting a more reserved stance within NATO, or even by opting for a policy of non-alignment; by restricting or denying the use of their territories to the Alliance, they would encourage rather than discourage Soviet aspirations and could run the risk of inviting Soviet pre-emptive acquisition. But the problems of security do not seem to attract much serious attention from the public in these communities. This is true, especially in the Faroes and in Greenland, but also to some extent in Norway. The exploitation of resources preoccupies people more, and may in the end influence public attitudes to security.

THE POLITICS OF RESOURCES

The ways in which problems concerning resources in the North Atlantic may affect security policies differ according to the kind of resources concerned. For example, protein resources could cause conflicts between one or several of the littoral communities and some of their NATO allies, which could strain relations within the Alliance. In the worst case they might cause one of the littoral communities to withdraw from NATO, or to limit participation in the Alliance to a minimum, excluding the maintenance of certain military facilities on her territory. In any event, such conflicts might play into the hands of the Soviet Union.

In a new Cod War, the littoral community concerned would find a lot of sympathy among the public in the other littoral communities, even in the Faroes, which unlike the others, is not primarily interested in an extension of the fishery limits. Within all the communities such conflicts could cause increased estrangement from the alliance, and even develop into non-alignment.

The discovery of hydrocarbon resources in the North Atlantic could produce a variety of effects on policies and decisions relating to security in the area. The possible effects depend, among other things, on the location and the quantity of such resources. In the North Atlantic region, oil and gas have so far not been found in economically profitable quantities outside the North Sea. But there are several prospective areas for offshore drilling. If oil or gas, promising huge revenues to the communities, are found off Greenland or the Faroes, one effect may be a demand for a constitutional change involving full or increased independence from Denmark. In the case of the Faroes, such a demand would be quite likely, but in the case of Greenland less so. With the prevailing neutralist attitudes of the Faroese such a development would certainly have an impact on security policy, and though it is less likely in Greenland in the short run, it would be of more far-reaching consequences if it were to occur. The discovery – and even the mere expectations – of striking oil or gas in the Barents Sea may have quite different effects on the security environment. If Norway is forced to comply with the Soviet position with respect to the division of the Barents Sea, a precedent could be set in Norwegian–Soviet relations which might affect the future pattern of security relations as well. Alternatively, the Norwegian authorities might be tempted to seek some kind of compromise with the Soviet Union within an emerging bilateral framework which could complicate Norway's position and in effect limit her role within the multilateral security framework. Structural changes in the North Atlantic region would then be likely to occur.

PHILLIP A. KARBER and JON L. LELLENBERG

The State and Future of U.S. Naval Forces in the North Atlantic

The purpose of this paper is to explore the state and future of United States naval forces in the North Atlantic. The approach it takes to the subject is a three-fold one. First, a survey has been conducted of the kinds of naval forces which the United States currently possesses and is continuing to develop. Second, an examination of the present deployment patterns of these naval forces has been undertaken on a global basis, in the Atlantic Ocean, and in the North Atlantic itself. In both cases a comparative method has been employed, so that the treatment of the Soviet Navy does not duplicate what can be found elsewhere in this volume, but instead puts the nature of and trends in United States naval forces into perspective. Third, the proposed employment of United States naval forces in the North Atlantic in both peacetime and wartime has been described according to current doctrine and planning assumptions.

On the basis of the preceding sections, the paper then turns to the identification and analysis of certain contradictions which may exist between the United States naval forces as currently deployed in the North Atlantic area and the political and military missions which they may be called upon to fulfill. The final section looks at future trends in United States naval forces in the Atlantic and North Atlantic, in terms of alternative force options over the next five to ten years.

UNITED STATES NAVAL FORCES TODAY

In recent years a frequent American approach to discussions of the threat at sea has been to emphasize the growth of the Soviet Navy, focusing upon aggregate numerical comparisons of its expanding size with that of the declining American fleet. But the United States and Soviet Navies have followed their own rather different paths of development over the past twenty years and more, reflecting considerable divergencies between the two nations on such influential aspects as perceptions of their adversaries, commitments and requirements, over-

30

Fig. 1 Comparison of U.S. and Soviet fleet size

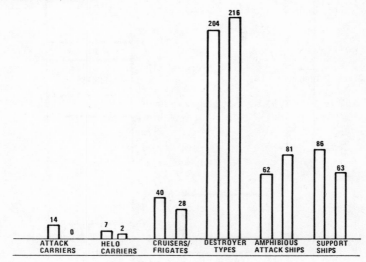

SURFACE SHIPS IN OPERATION 1974

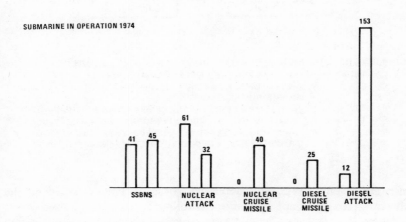

SUBMARINE IN OPERATION 1974

seas basing, institutional tendencies and prejudices, and budgets. Generally the number of ships in the US Navy has tended to decline over the last twenty years, particularly after 1968, when many of the oldest ships dating from World War II were finally decommissioned. Over the same period, the Soviet Navy underwent a great expansion

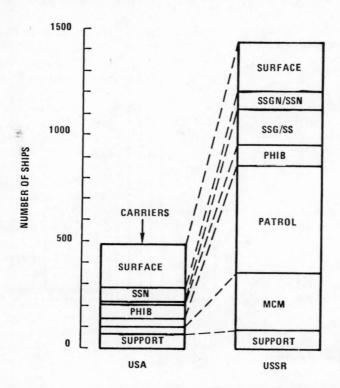

Fig. 2 Total number of ships

KEY: SURFACE = MAJOR SURFACE COMBATANTS (OVER 1,000 TONS)
SSGN/SSN = NUCLEAR POWERED ATTACK SUBMARINES
SSG/SS = DIESEL POWERED ATTACK SUBMARINES
PHIB = AMPHIBIOUS LIFT SHIPS
PATROL = PATROL VESSELS (SURFACE COMBATANTS 1,000 TONS AND SMALLER, NOT INCLUDING INSHORE/RIVER CRAFT)
MCM = MINE COUNTERMEASURES VESSELS (SEA GOING VESSELS ONLY)
SUPPORT = MAJOR SUPPORT AND UNDERWAY REPLENISHMENT SHIPS

from an essentially coastal defensive force to a major ocean-going navy with a significant offensive capability.

Figure 1 compares the numbers of major combatant and combat support ships in the United States and Soviet Navies during 1974, broken down in kinds of surface vessels and submarines.[1] The overall sizes of the surface fleets are not greatly different, with an edge of less than twenty-five units in favour of the United States, while the Soviet submarine fleet is more than three times the size of that of the United States, excluding strategic nuclear-powered ballistic submarines (SSBN) in both cases. Considerable differences in force design and structure are obvious from the numerical comparisons. On the United States side of the ledger there is her large and unrivalled aircraft carrier attack force, complemented by her current superiority in numbers of helicopter carriers. The considerable presence and continuing development of these two ship types is commensurate with United States history and emphasis on amphibious operations and projectable strike power. Compensating for these carrier forces, in the Soviet view, are the Soviet Navy's large numbers of cruise missile submarines with both nuclear and diesel power – – a force component designed primarily for defensive and anti-shipping operations, and of potentially great significance, especially if used in a pre-emptive manner. The US Navy maintains a significant advantage, quantitatively and also qualitatively, in nuclear attack submarines.

In most other types of major combatant surface vessels, amphibious attack ships, and support ships, the United States and Soviet Navies are fairly equal in numerical strength. Under current conditions it is the Soviet Navy's vastly greater numbers of patrol vessels of 1,000 tons or less displacement and mine countermeasure vessels that gives it an almost three-fold quantitative superiority over its United States counterpart. Figure 2 illustrates the contribution these two types of naval vessels make to overall Soviet fleet strength. While the value of these particular types of ship should not be ignored, they do not contribute greatly to the Soviet Union's status as a great oceanic naval power *vis-à-vis* the United States and her allies. But it should be noted where the important discrepancies do exist in comparisons of American and Soviet naval combat power: the United States can mount massive conventional and/or nuclear air campaigns launched from mobile platforms at sea, while the Soviet Navy is capable of conducting one of the potentially most effective submarine anti-shipping campaigns in the world, including – and perhaps especially – in the Atlantic and North Atlantic.

Figure 3 compares the major United States and Soviet combatant forces in terms of the average age of ships. The average age of American carriers, cruisers, and destroyers is greater than their Soviet counterparts, but less in the case of other ocean escort vessels and attack submarines. In the latter cases particularly, however, this somewhat younger age is largely the result of decommissioning many of the oldest vessels in the U.S. Navy over the past decade. This is particularly true of diesel attack submarines and large numbers of escort vessels dating from World War II. Twenty years ago there were over one hundred of these submarines in the US Navy; today

Fig. 3 Comparison of U.S./Soviet Average Age of Ships 1974

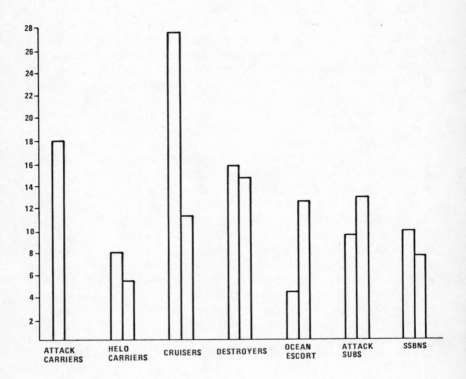

there are no more than a dozen left. The decline in the Soviet diesel submarine fleet since the end of World War II has been nowhere near as drastic, and took place largely among submarines of coastal and medium range. Twenty years ago the US Navy had over 350 escort vessels; in the following twelve years that number declined by only about 20 per cent, but subsequently dropped a precipitous additional 40 per cent in the next six years through decommissions.

So the present average age of United States naval vessels is as young as it appears largely through the recent phasing out of elderly and obsolescent vessels, while in general the modern Soviet Navy dates from the late 1950's, after Admiral Sergei M. Gorshkov became its Commander-in-Chief and initiated the large-scale modernization and expansion programmes that have resulted in today's Soviet fleets. The relatively younger average age of certain types of Soviet vessels does not, however, mean that they are necessarily qualitatively superior

34

Fig. 4 Comparison of U.S./Soviet Ship Construction 1965–1980

BOTTOM BAR 65 - 74
TOP BAR 75 - 80

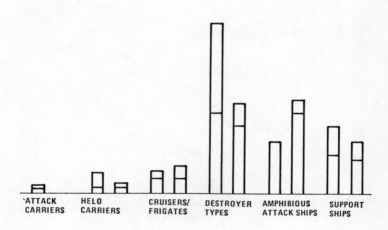

SURFACE SHIPS

'ATTACK CARRIERS HELO CARRIERS CRUISERS/ FRIGATES DESTROYER TYPES AMPHIBIOUS ATTACK SHIPS SUPPORT SHIPS

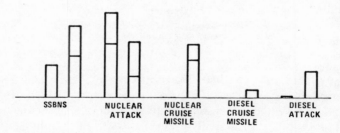

SUBMARINES

SSBNS NUCLEAR ATTACK NUCLEAR CRUISE MISSILE DIESEL CRUISE MISSILE DIESEL ATTACK

to their American counterparts. In certain technological areas the United States is clearly ahead, for example in silencing and sonar capabilities, through the Soviet Union has gone further in arming her combat vessels with guided missiles and cruise missiles.

Figure 4 compares United States and Soviet ship construction pat-

35

Fig. 5 Comparison of U.S./Soviet Fleet Tonnage

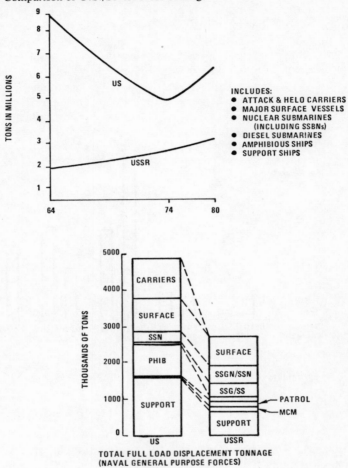

INCLUDES:
- ATTACK & HELO CARRIERS
- MAJOR SURFACE VESSELS
- NUCLEAR SUBMARINES (INCLUDING SSBNs)
- DIESEL SUBMARINES
- AMPHIBIOUS SHIPS
- SUPPORT SHIPS

TOTAL FULL LOAD DISPLACEMENT TONNAGE
(NAVAL GENERAL PURPOSE FORCES)

terns from 1965 to 1980. The bottom bar in each case signifies the period from 1965 to 1974; the top bar from 1975 to 1980. In the Soviet case the latter represents projections of Soviet naval construction based principally upon the extent of current and continuing naval shipbuilding programmes. On the American side, growth in this period has been most notable in destroyers, nuclear attack submarines and support ships. Among frigates and destroyers there has been a gradual decline in gun-armed types, with guided missile-armed frigates and destroyers assuming an increasing percentage of the total.

36

No new diesel attack submarine construction is planned by the United States, and the decommission of existing units is programmed to continue. Among Soviet naval vessels, on the other hand, further growth in helicopter carriers, cruisers, frigates and destroyers is foreseen, with guided missiles likely to receive major emphasis for primary armament. Nuclear power is probable as the basis for continued expansion of the attack and cruise missile submarine fleets. In other words, no drastic changes appear likely to occur in either navy over the next five year period. The United States plans to rely upon her now traditional carrier-oriented surface task forces and nuclear attack submarines, while the Soviet Union proceeds with her development of naval forces geared to high political visibility and presence in peacetime, sea denial and deterrence in time of crisis, and pre-emptive, attritive operations in wartime.

Finally, Figure 5 summarizes the overall trend in United States and Soviet naval growth by indicating comparative total tonnages. Here the Soviet Navy's great number of light patrol and mine countermeasures vessels does little to compensate for the US Navy's great investment in her attack carrier forces. Since 1964 the overall tonnage of United States naval forces had declined, principally through decommissions, reaching its low point with the completion of this programme around 1974, and it has now begun to rise again slowly with new naval shipbuilding underway. The Soviet Navy has increased gradually throughout this period of time, and should continue to do so through 1980, as current naval construction programmes there go forward.

UNITED STATES AND SOVIET NAVAL DEPLOYMENT IN THE NORTH ATLANTIC

There are still some observers in the United States who believe that the oceans in general, and the North Atlantic in particular, are relatively unimportant to the Soviet Union as a supposedly self-sufficient land power. In fact, the Atlantic and North Atlantic waters are of considerable importance to the Soviet Union, embracing as they do a set of regional, strategic, and general considerations that have great significance for both superpowers.

For the Soviet Union it begins as a simple question of the region's geography. As long ago as World War II the Soviet Union took pains to make this interest clear, when Molotov told Norway's Foreign Minister Trygve Lie:

'. . . the Dardanelles . . . here we are locked . . . . Oresund . . . here we are locked in. Only in the North is there an opening, but this war has shown that the supply line to Northern Russia can be cut or interfered with. This shall not be repeated in the future. We have invested much in this part of the Soviet Union, and it is so important for the entire Union's existence that we shall in the future ensure that Northern Russia is permitted to live in security and peace.'[2]

This human and economic investment in the Murmansk area and the Kola Peninsula as a whole has been massive, in terms of directed colonization and determined development of its mineral, raw material, and fishing resources. The military development of the Kola Peninsula has more than kept pace, and today the Kola Peninsula is one of the most highly-militarized areas in the entire Union. At the end of World War II the North Fleet was the smallest in the Soviet Navy. Today it is the largest, with some five hundred vessels, several hundred aircraft, and approximately 75 per cent of the Soviet Union's modern SSBN force – the Y-class and D-class submarines with their SS–N–6 and SS–N–8 SLBM.

Only the North Fleet enjoys relatively open access to the Atlantic Ocean. The Baltic and Black Sea Fleets remain essentially local forces by comparison, despite the Mediterranean activities of the latter. But even the North Fleet's access to the Atlantic cannot be taken for granted by the Soviet Union in time of crisis or war. Given the climatic conditions of the far north, especially during the winter months, the Barents and Norwegian Seas represent a great fjord, offering access for the Soviet naval forces based at Murmansk to the open waters of the Atlantic Ocean. From the Soviet viewpoint, the control of this fjord in the past has unfortunately been largely in the hands of its probable adversaries, Norway, Britain and America. More recently, however, the growth of Soviet naval power has enabled the North Fleet to extend its frequent presence far to the west of the North Cape, and Soviet naval exercises conducted since 1968 have suggested that the Soviet naval defensive zone now begins along the line running from Greenland through Iceland and the Faroes to Norway; that is, at the mouth of that great fjord opening into the Atlantic.

Moreover, control of these waters would be necessary to help defend against the threat posed by the Polaris and Poseidon submarine-launched ballistic missiles (SLBM) carried on American and British submarines. The ranges of these missiles are such that satisfactory coverage of major urban-industrial targets in the Soviet Union requires the SSBN to operate in relatively close proximity to the Soviet coastline, particularly in the far north. However, the operational capability in several years of the new Trident system will allow much the same SLBM target coverage over considerably greater ranges, and in this sense the North Atlantic may actually be of declining interest to the United States. Yet, despite this declining value for United States SSBN deployment, freedom of operation in these North Atlantic waters remains of significance for the Soviet Navy.

The bulk of its SSBN fleet operates out of Murmansk, and it is consequently in the Norwegian Sea and the North Atlantic that enemy anti-submarine warfare forces would have the greatest likelihood of detecting and destroying Soviet Y-class and D-class submarines.

Finally, the Atlantic and North Atlantic are of great significance to the Soviet Union because of the possibility of war in Europe. United States wartime employment of her naval forces in the region is envisaged mainly in the context of a protracted armed conflict in Europe, and focuses principally upon the reinforcement and re-supply of United States and allied forces deployed there and engaged in combat operations. Some 90 per cent of these fresh troops and additional military supplies and equipment must be brought from the United States by ship, and therefore it would be of the utmost importance to the Soviet Union in these circumstances to prohibit the sea lanes in question and as far as possible to prevent American reinforcement and re-supply from taking place. For greatest effectiveness it must do this rapidly and with high levels of successful transit, placing no small value upon Soviet naval readiness and presence in these waters in peacetime.

So the North Atlantic is an area of increasingly vital interest to the Soviet Union. The structure and orientation of Soviet naval forces have changed over the last fifteen years from coastal defence to blue-water offence. Nevertheless, United States naval force deployment appears oblivious to this growth in both the North Atlantic interests and the capabilities of the Soviet Navy.

By most standards of measurement Europe easily ranks highest in United States interests, with Asia and the Middle East tying at present for a considerably lower second place and the rest of the world lower still. It is in Europe that large standing adversary army and air forces are arrayed opposite the largest number of overseas-deployed United States troops, and where a large number of Soviet naval forces are concentrated in the North and Baltic Fleets. Similarly, it is in Europe that the danger of the introduction of nuclear weapons in a conflict, with its attendant potential for escalation representing a threat to millions in the United States, is highest. Again, there is at present only a relatively secondary threat of a similar nature in Korea and the Middle East.

How does the deployment of United States surface and submarine

Fig. 6. U.S./USSR COMBATANT DEPLOYMENTS*

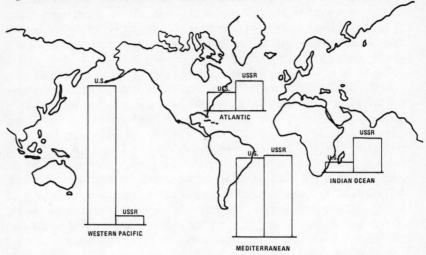

* Includes aircraft carriers, general purpose submarines, major surface combatants, minor surface combatants, amphibious ships, and mine warfare ships.

naval power reflect this? Figure 6 illustrates the relative strength of US naval deployments by oceanic region: the highest by far is in the Pacific Ocean, the second highest in the Mediterranean, and naval forces in the Atlantic Ocean make a not-terribly-close third. The United States has apparently deployed its naval forces in inverse relation to its scale of national interests and to the perceived threat. Nonetheless, it is not difficult to explain this. Until relatively recently the North Atlantic has been an Anglo-American lake. However, the very success of this Anglo-American predominance in the Atlantic Ocean has produced an eviscerating form of benign neglect.

The Pacific Ocean, on the other hand, has been the prime focal point for the US Navy for over a hundred years. During World War II the Navy's major offensive contributions to victory (as distinguished from defending the sea lanes of the Atlantic) were in the Pacific. It was a naval attack at Pearl Harbor which brought the United States into the war, and the culmination of a three-year naval offensive against Japan which ended that war. Likewise, the only two major United States post-war conflicts have been in that general area – in Korea and Vietnam. And the United States continues to have major commitments and interests in the Far East which depend to a great extent upon continued strong naval forces in the Pacific Ocean.

40

Fig. 7 Comparison of United States/Soviet Atlantic-Oriented Fleet Structure 1974

	US 2nd Fleet Atlantic	USSR North Fleet
Carriers (Attact & Helo)	4	0
Surface Combatants	67	56
	71	56
Submarines		
Fleet Ballistic Missile		
Submarines*	28	50
Cruise Missiles	0	40
Attack	20	80
	48	170
Total	119	226

* Includes all United States ballistic missile submarines based in Charleston, Holy Loch and Rota.

Figure 7 compares the assigned strength of the United States Second Fleet to the Soviet Union's North Fleet, operating out of Murmansk. The numbers suggest an American edge in surface vessels and a major Soviet advantage in cruise missile and attack submarines. Yet these numbers are somewhat misleading. Many of the United States ships in question are not actually of operational status; during 1975, for example, US naval forces assigned to the Atlantic area were comprised of approximately three to four carriers, three to four cruisers, fifteen to twenty destroyers, eight frigates, and seven to ten nuclear attack submarines. On the whole, this is less than 25 per cent of overall United States major combatant vessel strength. Naval air combat strength in the Atlantic was correspondingly limited: approximately five F-4 and 6 A–6/A–7 fighter squadrons, two E–2 early warning squadrons, two A–6 electronic warfare squadrons, and one A–5 reconnaissance attack squadron.

These are relatively meagre forces if compared to the naval strength available to the US Third and Seventh Fleets in the Pacific Ocean, or even the Sixth Fleet in the Mediterranean. But they must none the less stretch themselves to cover a rather extensive geographic area. The limited resources of the Second Fleet are responsible for United States naval presence and activities not only in the Atlantic Ocean proper, but in the North Atlantic, South Atlantic and Caribbean Sea as well. Thus a given 'snap-shot' day during 1975 might find approximately two carriers, four cruisers, a dozen destroyers and frigates, and perhaps half a dozen nuclear attack submarines operating in the

Atlantic area itself, with perhaps two destroyers and two nuclear attack submarines assigned from these forces to the Caribbean and the South Atlantic, and another carrier to the North Atlantic.

Figure 8 presents a comparison, during the 1964–1975 period, of United States and Soviet shipdays in the Atlantic Ocean. Shipdays are not an altogether meaningful standard of measurement in themselves, but they nevertheless do give some basis for comparing presence and activity. Soviet shipdays in the Atlantic have generally been on the rise throughout this period, particularly since 1968, peaking later at the time of the 'Okean' excercise. Since then there has been a fairly steady if somewhat lower annual rate, which is none the less vastly greater than that of a decade ago. United States shipdays per year in the Atlantic, on the other hand, have generally been on the decline during the same period. In 1974, Soviet shipdays in the Atlantic totalled some 15,000, compared to about 13,000 for the US Navy. Again it must be pointed out that these are not entirely straightforward or meaningful figures. If a portion of the Soviet Navy's shipdays are due to a surge factor rather than continual presence, the same tends to be true of the United States Navy as well. Of the four currently active fleets in the US Navy, considering only those vessels of operational status, the Second Fleet in the Atlantic Ocean has the lowest level of readiness. Many units, from nuclear attack submarines and destroyers to the aircraft carriers themselves, are barely prepared for combat operations.

How long would be required to generate a high state of readiness is not clear.

United States naval presence and activity consists of one aircraft carrier (the *Kennedy,* programmed for about fifty A–6/A–7 and twenty-five F–14 fighter aircraft, ten anti-submarine aircraft, and a dozen early warning and electronic warfare aircraft), one frigate, one cruiser, and one nuclear attack submarine. It would be unlikely to find all four vessels operating in the North Atlantic at the same time under normal conditions. In addition, there are approximately 3,000 US Air Force and Navy personnel stationed in Iceland, responsible principally for an F–4 interceptor squadron and about a dozen P–3 patrol aircraft. This scarcely amounts to a large naval force for a maritime region of growing political, economic, and military importance to the United States and its allies in NATO.

Under proper conditions and with sufficient warning, of course, it might prove possible to augment these forces considerably. In September 1972 NATO conducted its 'Strong Express' exercise with a dozen member nations contributing some 65,000 men, 350 ships, and 700 aircraft to the relief of a Norwegian ally under simulated attack by

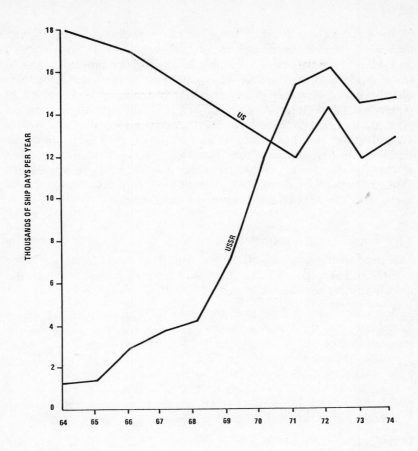

FIG 8 COMPARISON OF U.S./SOVIET SHIP DAYS IN THE ATLANTIC 1964 - 1974

the Soviet Union. ACE (Allied Command Europe) Mobile forces reinforced the Norwegian defence, and were themselves reinforced by additional landings of American, British and Dutch Marines supported by fighter and bomber aircraft cover from the *Kennedy* and the Royal Navy's venerable *HMS Ark Royal*. A *Kennedy*-led task force landed the marines, while other units of the American, British, Canadian, Norwegian, Danish, Dutch, and German navies participated in anti-submarine warfare operations, air-strike support to the landings, convoy operations, and minelaying and minesweeping operations in the Atlantic Ocean and the Norwegian, North, and Baltic Seas.

The exercise was conducted against fairly modest simulated Soviet air and naval resistance, and its conduct suggests that United States and Allied reinforcements available to augment committed forces in this region could require at least a week to arrive upon the scene. Nor did the exercise result in any increase in the level and volume of American naval presence and activity in the North Atlantic during normal conditions. There have been proposals recently for a greater United States naval presence in the North Atlantic, for example detailing to this region one of the Sixth Fleet's carriers usually stationed in the Mediterranean, but whether there will actually be any significant expansion of United States naval missions, forces, and activities there over the next five years remains uncertain.

THE RANGE OF MISSIONS

The general military objectives for American naval forces embrace a fairly broad range of general-purpose force missions, of which the principal ones are the following:

- defence of sea and air approaches to the United States, Western Europe, and the Panama Canal;
- maintenance of the security of the key United States island bases in the Atlantic Ocean area;
- military operations (especially anti-submarine warfare) to defend the United States from strategic nuclear attack; and
- defence of sea and air lines of communication and supply between the United States and Western Europe.

In peacetime the employment of US naval forces in the Atlantic and the North Atlantic is designed to support the successful achievement of these objectives in time of crisis or war. This is to be accomplished by a number of means: by establishing and maintaining a state of readiness to conduct combat operations successfully against the Soviet air and naval threat in the area, in the event of hostilities; by contributing to reconnaissance and early warning of attack (either by Soviet general-purpose forces or strategic nuclear forces at sea); and continual efforts at submarine detection and location at key choke points in these waters. In addition, United States naval forces have as a basic and traditional peacetime mission the maintenance of a visible and effective presence in order to support American national interests and serve United States foreign policy.

In wartime, however, the US Navy as a whole assumes two basic missions. The first is to secure control of sea and air lines of communication, in order to reinforce and re-supply other United States and

Allied military forces which are deployed overseas and engaged in combat operations. The second is to project additional United States military force into land combat areas by means of carrier-based naval aviation, gun and missile weaponry on board ships, and amphibious combat forces and operations. Both missions have been very familiar ones for the US Navy since World War II. But today the emphasis for the United States Second Fleet in the Atlantic Ocean area is placed heavily upon the first of these two missions. The wartime employment of US naval forces in the Atlantic Ocean and in the North Atlantic is seen mainly in the context of a major armed conflict that has broken out in Europe and which depends principally for its outcome (barring escalation to the level of general war between the two superpowers) upon the contest between NATO and Warsaw Pact ground and air forces in Central Europe.

In such circumstances US naval forces, together with those of NATO allies, would seek to counter the Soviet attack and cruise missile submarine force operating in the Atlantic and North Atlantic, defend and reinforce the Alliance's important strategic island bases in the area, and control the sea lanes necessary for reinforcement and resupply. In addition, United States anti-submarine warfare forces would seek to render ineffective the strategic threat to the United States comprised by the Soviet Navy's SSBN operating out of Murmansk.

It seems fairly unlikely that all of these assignments could be undertaken satisfactorily without considerable transfer of United States naval assets currently assigned to other fleets and commands. At best that would consume precious time, and at worst naval conditions in other regions, particularly the Mediterranean, might make it a very hazardous action politically and militarily. As a result, the US Navy has recognized that the reduced naval forces available to it in the Atlantic and North Atlantic require it to set priorities among the various assignments which they would be called upon to undertake.

Its first priority has been defined as securing control of the sea and air lines of communication that link the United States and Western Europe across the Atlantic Ocean. This decision is the result of current United States and NATO doctrine and planning assumptions for a major European armed conflict. These place heavy stress upon the need for the United States to convey reinforcements and supplies to embattled American and Allied ground and air forces on the European continent itself, engaged in combat operations which it is thought would be likely to continue for as much as several months before a termination of hostilities would be achieved.

The earliest part of this reinforcement and re-supplying effort will take place through airlifts from the United States, but it is also recognized that the great majority of troops and especially the many thousands of tons of supplies and military equipment – probably at least 90 per cent of them – would have to move by ship. In order to accomplish this, particularly if the Soviet Union exerts herself to interdict United States reinforcement activities and has been able to deploy her submarine force in advance, the US Navy expects to concentrate its limited forces in the Atlantic and North Atlantic upon four major tasks. First, United States naval forces will seek out, locate, and destroy Soviet naval forces in these waters, especially Soviet submarines. Second, they will prevent use of the air and sea lanes from the Soviet Navy's North Fleet bases into the Atlantic area critical to United States reinforcement and re-supply shipping. Third, they will reinforce United States naval and air forces based on important strategic islands, notably Iceland. And fourth, they will extend close support to shipping crucial to the reinforcement and re-supply effort, especially in the earliest days of the conflict, before the Soviet submarine threat is brought under control.

These are certainly rather challenging tasks which will stretch the limited United States naval forces currently available for duty in the Atlantic and North Atlantic to the utmost, and tax their present capabilities severely. Even so, a number of other desirable missions for United States naval forces, which could conceivably make important contributions to the conduct and outcome of the war as a whole, will probably go undone.

POTENTIAL PROBLEMS

Given their limited resources while facing a growing Soviet Navy, United States naval forces in the North Atlantic are thus being asked to do more with less. There is an inherent danger in asking too much – – the US Navy may simply not be able to do it. In the event of a European conflict, the US Navy will probably be forced to choose between two general-purpose force missions: projection of naval power into Central Europe in support of NATO's land forces, or securing control of the North Atlantic lines of communication.

Long War vs. Short War

In 1967 NATO adopted the strategy of 'flexible response', placing strong emphasis upon the forward deployment of sufficient ground forces so as to offer the option of an initial conventional defence. It was clearly stated that flexibility was needed to offer a chance of a negotiated ceasefire before escalation, but should NATO's defence be overrun or the Warsaw Pact continue to press the attack, nuclear weapons would be used. This is still the case today.

But given the structure and deployment of NATO's ground forces, they will have a difficult time surviving even a short war. They are outnumbered, thinly deployed, and lack the necessary depth to trade space for time. The European members of NATO clearly do not plan on a long-war strategy, and in fact maintain such limited stocks of war materiel that they would probably run out of supplies within thirty days. Lack of standardization of equipment among the Allies means that most United States logistic support would be unusable. For a campaign longer than thirty days, over 90 per cent of the materiel coming from the United States would have to come by ship. Even assuming no naval interference, the time required to mobilize, concentrate, embark, ship, unload, and move materiel to the forward area could require at least a month for the majority of equipment.

The current emphasis on United States general-purpose naval forces in the North Atlantic is based upon a long-war hypothesis and the need to conduct a massive anti-submarine warfare campaign. With a focus upon the need to apply all available conventional capabilities toward the initial stages of a short war in Central Europe, priorities in naval missions may need to be modified.

Tactical Nuclear Weapons

A second problem with the traditional long-war assumptions is that the longer the conflict lasts, the greater the probability of nuclear weapons being introduced.

There are over ten thousand nuclear weapons currently deployed in Central Europe. They are not solely under American and Soviet centralized control; Great Britain and France also have their own national arsenals. If the conventional defences begin to crumble, NATO plans to use nuclear weapons to bolster them. If the conventional attack becomes bogged down, Warsaw Pact strategy calls for nuclear fire-power to knock holes in NATO's defences. Even if neither side really wants to use nuclear weapons, given the scope, intensity and magnitude of a Central European war, the chances of nuclear escalation and accidental use grow with each day of conflict.

There is, of course, some chance that the employment of tactical nuclear weapons might remain within the geographic boundaries of continental Europe, but the prospect is slight if for no other reason than the orientation of naval forces. In fact, most anti-submarine weapons not only have a nuclear capability, but many are only effective when actually employing nuclear warheads. Yet support facilities are fixed targets and surface platforms are very visible and vulnerable.

Recent analysis thus leads to the conclusion that:

'nuclear war at sea would have very little impact on the survivability of ballistic missile submarines or of attack submarines, but
it would undoubtedly spell the end of the relatively few capital
surface ships . . . Nuclear war confined to sea appears to degrade
ASW capability far more than it degrades the survivability of the
submarine.'[3]

Should a European conflict escalate to involve nuclear weapons at
sea, the Soviet Navy, with its emphasis upon submarines, cruise missiles and sanctuary-based long-range patrol aircraft, is likely to have
the edge. Similarly, while the probability of convoy survival in a nuclear
environment is very poor, it is none the less greater than the probability
that there will be a functioning port at which to disembark. Thus, an
early NATO initiation of nuclear weapons to stem the attack and buy
time for reinforcements might actually decrease the probability of
receiving them.

The waging of tactical nuclear warfare at sea has several potential
advantages over a similar level of conflict on land: military targets are
easier to acquire; collateral civilian damage must be vastly lower; and
the sea forms a clearly perceptible and delineated zone of engagement. Yet NATO can hardly rejoice at the prospect. Tactical nuclear
warfare at sea would probably favour the Soviet Union. NATO's initiation of nuclear weapons on land in Central Europe could spill over
to the sea, and thus destroy the prospect of reinforcement. The
chances of accident, miscalculation, or a rushed judgment are actually
higher with naval forces.

Nor is it certain that use of nuclear weapons could be confined to
the sea. While the Atlantic has clearly perceptible geographic boundaries, these could seem like incentives, rather than inhibitions, for
escalation. There are many lucrative targets on the borderline, such
as shore-based airfields and communications facilities supporting
naval activities, large base complexes and key ports. In a nuclear campaign at sea these would soon become prime targets. The Atlantic is
NATO's rear area and the use of nuclear weapons there, like strikes
deep in Eastern Europe, could quickly lead to nuclear use on the
central front. The Atlantic littoral also borders both superpowers,
thus increasing the prospects of escalation towards direct confrontation. Even worse, the Atlantic serves as the deployment zone of the
key SLBM element of both superpowers' strategic deterrent. Both
conventional and tactical nuclear ASW campaigns have the awesome
potential of disturbing the most stable element of the strategic balance,
the consequences of which should not be minimized in any optimistic
discussion of a long war in Europe.

Sea Control or Support of the Central Front?

Traditionally, the mission emphasis of United States naval forces in the Atlantic and North Atlantic has been neither the conduct of Navy-initiated offensive operations nor the integration of naval combat power with support of United States ground and air forces. Instead, the power has been relegated to the tertiary role of keeping the sea lanes open to allow reinforcement and re-supply to Europe. This is an historical legacy from our experience in two World Wars and it may no longer be appropriate for the military and political conditions which exist in Europe today.

There is a pervasive predisposition among Western national security observers to view today's Soviet threat to Europe in past terms, to view the Soviet Navy as being analogous in operation to the submarine fleets of Imperial Germany and the Third Reich. But there is a danger in historicism. What was past may be prologue, but it is not necessarily predictive. For example, the Soviet Union may not act like the Germans; the conflict may not be a long conventional war, but instead be short and perhaps nuclear. Costly mission priorities are not absolutes and may change as the conflict changes.

A protracted submarine campaign against Allied shipping in the event of a European conflict may not have the priority the West thinks it has. If the Soviet Union believes what it says about a European war being short and nuclear, a massive submarine campaign in the Atlantic would be superfluous.

'The Soviet Union seems to have configured its ASW operations and other naval forces towards the possibility of a nuclear confrontation. The stated tasks of the Soviet naval forces are: to prevent US carriers from launching their nuclear-armed aircraft from the Norwegian Sea and Eastern Mediterranean; to provide an effective counter to the US *Polaris* fleet; to defend Soviet surface ships from US submarines; and to protect their own ballistic missile-carrying submarines. The Soviet hunter-killer submarines seem to be designed and used to support and protect forward-deployed Soviet surface units and ballistic submarines rather than to attack merchant shipping in the North Atlantic or elsewhere.'[4]

The limited Soviet deployment of attack submarines in the North Atlantic during peacetime and the focusing of their exercises in the Norwegian Sea tend to lend credence to this interpretation of Soviet doctrine.

A second drawback in depending upon historical parallels is the failure to appreciate the differences between World War II and a future war in Europe. Once commandeered into national service and loaded for shipment, a process which would take from one to two months, the United States merchant shipping tonnage available would far exceed both what was used in World War II and what is required for re-supply to Europe. Today's merchant ships are faster, more sophisticated, and have much greater tonnage carrying capacity than their World War II counterparts. In addition, the point defence capabilities of today's convoy-protection ASW systems are very much greater than those of World War II. Compared to thirty years ago, the submarine versus merchant convoy balance has shifted dramatically in favour of the convoy.

So there appears to be what might be called the 'North Atlantic paradox' in present United States naval strategy:

- Transatlantic mobilization and re-supply, if initiated several months before a conflict, could drastically alter the conventional inferiority of NATO in Europe.
- In a short, intense conflict with little lead warning time, however, the current US mission emphasis upon sea control will contribute little to NATO's overall defence.
- A prolonged conflict is likely to become nuclear, in which case surface lines of communication and shipping will probably be lost.
- The current emphasis upon sea control may actually increase the probability of escalation to nuclear warfare.
- In the unlikely event that a European land war turns into a protracted conventional conflict, convoy survival ability will be high and protection can be provided with mobilized US-based naval support.

So the key question about the future of United States naval forces in the Atlantic and North Atlantic is whether, in the event of a European war for which NATO has not yet mobilized, the forward-deployed North Atlantic units of the US Second Fleet should continue to be oriented to the sea control mission, or whether this should be given secondary priority in favour of other naval missions more directly affecting the fighting on NATO's Central Front.

NEW OPTIONS

In 1975 the United States military, at the instigation of the then Secretary of Defense James R. Schlesinger, began to evaluate a number of options to expand the mission orientation of United States naval forces in the North Atlantic in this direction. The purpose was to provide a more direct contribution to the ability of the ground forces to survive the initial stages of Soviet conventional short-war strategy. Three basic functions could be provided by forward-deployed naval forces if they were not preoccupied with the long-war sea control mission, at least in the early stages of a European conflict:

- Use of naval nuclear assets, the *Poseidon* SLBM in particular, to free European-based tactical air resources for conventional combat;
- Use of naval carrier-based aircraft to supplement European tactical air resources; and
- Use of naval infantry (the US Marine Corps) to provide additional land combat capability in the Central Front.

While these three options have all been raised before elsewhere, many observers have expressed surprise and doubts about their practicality and acceptability. Some express doubts about using SLBM against theatre nuclear targets, yet the proposal has been the subject of several international conferences and published works since 1974,[5] and was discussed in a United States Department of Defense report last year.[6] Others doubt that the United States would be willing to shift carrier assets, yet in autumn 1975 the American press reported that Schlesinger had recommended part-time transfer of one Sixth Fleet attack carrier from deployment in the Mediterranean to patrol in the North Atlantic.[7] The most sceptical comments are aimed at the option of deploying United States Marine Corps units in Central Europe, yet in autumn 1975 Marine units actually conducted combat exercises in Northern Germany, for the first time in fifty-eight years. Since then a major Brookings Institution study has proposed the same option.[8] These options are controversial; there are many impediments to their implementation, and they may never be fully exercised. On the other hand, they are being discussed in the United States as measures to increase NATO's short-war capability,[9] and would make a significant impact on the state and future of United States naval forces in the North Atlantic.

Strategic Weapons in Support of the European Theatre?

The United States ballistic missile submarine fleet could be able to take over a growing share of the targeting for NATO's nuclear strike requirements in the future. Improvements in both nuclear warhead and guidance-system technologies are combining to provide SLBM with the yields and accuracies necessary to undertake precise attacks upon fixed Soviet/Warsaw Pact military objectives with minimal amounts of collateral damage. To do so could have severe disadvantages for strategic deterrence, but be important for theatre defence. It would reduce the present vulnerability of NATO's nuclear strike capability to a sudden Soviet pre-emptive attack at the outset of a conflict. It would free badly needed NATO tactical aircraft for both conventional and nuclear support of the battlefield ground forces. And it might reinforce the eroded deterrence link of the United States strategic forces to NATO Europe's defence, which sustains both the credibility of the overall deterrent posture and the political cohesion of the Alliance.

Naval Carrier-Based Aircraft to Reinforce European Aircraft?

The most pressing short-war requirement for United States naval forces in the North Atlantic is to provide support to the Central Front in the initial stages of a Soviet conventional attack. This can best be achieved not by trying to clear the Atlantic Ocean of Soviet submarines in anticipation of massive reinforcement convoys thirty days or more after the war has started, but rather by adding additional American carriers to operate close in to the continent and make a possibly decisive contribution to the struggle as a whole. Even a limited number of naval aircraft, particularly the high-performance F–14 and F–4, if deployed within range of the Central Front, could offer substantial assistance in meeting and defeating the Warsaw Pact air threat, thereby freeing crucial numbers of ground-based aircraft in the Central Front for use in the ground support role.

Providing this capability, however, requires a significant shift in emphasis for United States surface naval forces. First, naval air would have to exercise and plan for their use in the Central Region much more than they do today.

Second, the United States would probably need more carriers than would normally be available, given the load level of current American carrier deployment in the North Atlantic. Future emphasis may be given to stationing additional carriers and their complementary task forces in the North Atlantic area.

Third, the United States Navy, in close co-operation with the NATO allies, could maintain a higher presence than today in the North Sea, the Norwegian Sea and the English Channel as part of United States peacetime deployments. This shift in US naval forces would signify to the NATO allies and to the Soviet Union that the United States is not only committed to Europe but will react strongly to increased Soviet ground capabilities and naval presence.

And fourth, major United States redeployment of naval forces to Europe would not only have deterrent and defence advantages for the Central Front, but would also provide vital back-up for both NATO's northern and southern flanks. Thus, should the Soviet Union put pressure on Norway, they could be deployed quickly to the Norwegian Sea. Should another crisis break out in the Middle East, these forces could provide valuable support to the Sixth Fleet's operations in the Mediterranean.

United States Marine Units on the Central Front?

A third possible naval force realignment would be the forward deployment of United States Marine Corps units in Northern Germany. An examination of the current deployment of NATO ground forces in Central Europe shows that the American divisions are located in southern Germany but their main lines of communication and supply run on a north-south axis from Bremerhaven. It should be kept in mind that it is in northern Germany that the most likely Soviet penetrations would occur in a short and highly intense campaign, due to the open, rolling terrain of the North German Plain and the lack of adequate NATO ground forces deployed close to the border.

By basing a Marine Corps brigade around Bremerhaven, the United States could not only protect her own lines of communication, but also bolster NATO's NORTHAG (Northern Army Group) defences. This Marine brigade would obviously require a different Table of Organization and Equipment (TO & E). Light infantry is of relatively little use against a massive tank attack. However, if this Marine brigade were extensively equipped with the latest anti-tank technology, it could make a significant contribution.

CONCLUSIONS

The current state of United States Naval Forces in the North Atlantic is ambiguous. American interests in the area have not declined but her capacity to protect them from a growing threat has. Given the limited assets deployed in the North Atlantic, the United States is asking a lot of its Navy; perhaps too much.

The peacetime presence of the US Navy in the North Atlantic provides a symbolic manifestation of the American commitment to Europe. However, that presence is small and with modest effort the Soviet Navy can make it appear of diminished stature.

NATO has demonstrated a surge capability in support of the Northern flank should limited aggression occur. However, in the event of a concurrent Central European war, the Northern flank is likely to be on its own.

The military contribution of the United States Atlantic Fleet to the defense of Central Europe once a conflict has started is more dubious.

There are potential contradictions between the strategy of NATO's ground and tactical air forces and the traditional assumptions governing the employment of the United States Second Fleet, particularly in the areas of mission priority, the impact of tactical nuclear weapons, and in regard to the duration and intensity of conflict. Unless these problem areas are dealt with the future state of United States naval forces in the North Atlantic is not bright.

ROBERT G. WEINLAND

The State and Future of the Soviet Navy in the North Atlantic

INTRODUCTION

The objectives of this discussion are limited, and so is the range of subjects addressed. There are four parts.

The first part is a summary listing of the forces currently assigned to the Soviet Northern and Baltic Fleets and description of their operations in the North Atlantic. The second part is an examination of the likely employment of these forces – differentiating 'strategic offensive' from 'general purpose' forces, and addressing their respective missions and expected *modus operandi* in both peacetime and wartime. In the third part, two important questions are given closer consideration: the wartime requirement for these forces to have access to the Atlantic, and their use in peacetime as instruments of Soviet foreign policy. The last part of the discussion is an attempt to identify the implications of what has been said in terms of actions the Russians can be expected to take with respect to the Navy and to the North Atlantic.

Throughout, an effort has been made to avoid *over*estimating Soviet naval capabilities and the Russain intent to use them in ways inimical to Western interests. Alarmism, in the long run, is poor policy! An equal effort has been made to avoid *under*estimating those capabilities and intentions. These are potent forces, and they pose a clear danger to the West and its interests.

Purists will probably shudder at some of what follows. Among other things, this discussion is predicated on a purposefully indistinct conception of the geographical boundaries of the North Atlantic (the Baltic and Mediterranean are definitely excluded, but the other contiguous seas are not – most of the time). Western terminology is used in discussing Soviet forces, in particular the terms strategic offensive and general purpose forces. The former is a reference to the submarine-launched ballistic missile (SLBM) force; the latter refers most of the time to the remainder of the Navy. Finally, some of the numbers used here must be regarded as suspect; but they are all we have to work with, and the margin of error probably is not excessive.

FORCES, DEPLOYMENTS AND ACTIVITIES

Forces

This part of the discussion deals with those Soviet combat forces that can and do operate in the North Atlantic. In essence, this confines the discussion to the operational, ocean-going combatants and combat aircraft of the Northern and Baltic Fleets, and excludes their coastal defence, amphibious assault and auxiliary components.

There are two ways to describe the size and composition of these forces: one is in absolute terms – which provides a rough measure of their capabilities and the threat they pose – and the other is in terms of the section they represent of the entire Soviet naval inventory – which provides some insight into Soviet resource allocation priorities (and, indirectly, insight into these forces' *raison d'être*). Both approaches are employed here (see Table 1).

Table 1: Northern and Baltic Fleet Force Structure*

	Northern Fleet	Baltic Fleet	Regional Totals	Regional % of Soviet Navy Total
Submarines				
Ballistic missile	50–55	–	50–55	70
Cruise missile	40–45	—	40–45 }	60
Torpedo attack	80	35	115	
Total	175 (90 nuclear-powered)	35	210	60
Major surface combatants (Cruiser, destroyer, frigate types)				
Missile-equipped	15	15	30	50
Gun	40–45	40–45	85	50
Total	57.5	57.5	115	50
Combat aircraft				
Heavy bomber types	30	10	40	70
Medium bomber types				
Missile-equipped	150	75	225 }	70
Other	95	60	155	
Maritime patrol/ASW	65	45	110	50
Helicopters	80	55	135	
Total	420	245	665	60

* Data on submarines and surface combatants from: 1975–76 editions of *The Military Balance* and *Jane's Fighting Ships*. Data on combat aircraft from: Robert Berman, 'Soviet Naval Strength and Deployment,' in: MccGwire, Booth and McDonnell (eds.), *Soviet Naval Policy: Objectives and Constraints* (New York: Praeger, 1975), p. 423.

The Northern Fleet currently has approximately 175 submarines – 50–55 of which are ballistic missile-launching; 40–45 of which are cruise missile-launching; and 80 of which are torpedo attack types. Roughly half of all of these submarines are nuclear-powered. It has 55–60 major surface combatants (cruiser, destroyer and frigate types), 15 of which are equipped with surface-to-surface missiles (SSM) or surface-to-air missiles (SAM), or both. It also has some 340 fixed-wing combat aircraft and 80 helicopters.

The Northern Fleet is obviously submarine-orientated. Even if the strategic offensive component is subtracted from the submarine totals, submarines outnumber major surface combatants by two to one. More will be said about this below.

The Baltic Fleet has no strategic offensive component. It is a general purpose fleet. It is also more balanced than the Northern Fleet. It has 140 fewer submarines than the Northern Fleet – a total of 35, all conventionally-powered torpedo attack types. On the other hand, it has the same number of missile- and gun-equipped major surface combatants, 55–60. It has 190 fixed-wing combat aircraft – 75 of which are missile-equipped, half the number in the Northern Fleet – and it has 55 helicopters.

Aggregating these fleet inventories and calculating the relationship of the resulting regional totals to the total Soviet naval inventory gives a better picture of the threat they pose to NATO and its forces in the Atlantic. It also shows how the Russians think about this theatre.

The strategic offensive force consists of some 50–55 SLBM platforms, all based in the Northern Fleet. This is 70 per cent of the total Soviet SLBM force. The remainder is located in the Pacific. The general purpose submarine force totals some 155–160 units, one quarter of which are cruise missile-equipped. This represents 60 per cent of the Soviet Navy total. At any one time, however, roughly 10 per cent of the general purpose submarines based in the region are deployed to the Mediterranean, and are thus not available for North Atlantic operations. (The Montreux Convention, which governs passage of the Turkish Straits, prohibits Black Sea Fleet submarines from operating in the Mediterranean. Consequently, the submarine contingent of the Mediterranean Squadron must be drawn from the other Western fleet inventories. With few exceptions, they come from the Northern Fleet.)

Some 115 major surface combatants or half of the total Soviet inventory are based in this region; 30, or roughly one quarter of these, are missile-equipped. That also represents half of the total Soviet inventory of such units.

The regional inventory of naval-subordinated combat aircraft totals

665 units, and accounts for 70 per cent of the Soviet Navy's heavy bombers, 70 per cent of its medium bombers and 50–60 per cent of its fixed- and rotary-wing ASW aircraft.

Aggregating all of the Soviet Navy's anti-ship missile-launching platforms potentially available for North Atlantic operations subsurface, surface, and air gives a total of approximately 300 units. That is a formidable capability. In actuality, however, the figure is even higher, because missile-equipped elements of Soviet Long Range Aviation (LRA) routinely supplement units of the Soviet Naval Air Force (SNAF) conducting exercise strikes in the North Atlantic, and presumably would do so in wartime as well.[1]

Furthermore, while the exact nature and extent of its potential participation in such activity remains unknown, according to the late Soviet Minister of Defense, Marshal Grechko, some of the land-based long-range missiles of the Strategic Rocket Troops (SRT) apparently are (or can be) targeted against 'naval groupings in theatres of military operations (at) sea.'[2] A substantial portion of the North Atlantic is within range of a substantial portion of the SRT's missiles. That makes the strike capability that can be brought to bear in the North Atlantic even more formidable.

Before examining the operations of these forces, we should take a moment for a closer look at the Northern Fleet's submarine force, since it is quantitatively and qualitatively the most important component not only of that fleet but of the Soviet Navy as a whole (see Table 2). That was the case in 1960; it is even more so today.

Table 2: Soviet Submarine Force* (all types)

	Northern Fleet	Baltic Fleet	Black Sea Fleet	Pacific Fleet	Total Inventory
1960	140	85	65	120	430**
1965	150	70	50	120	410**
1970	150	75	40	105	370
1975	175	35	25	105	340

  * Data from 1960–61, 1965–66, 1970–71 and 1975–76 editions of *The Military Balance*.
 ** The differences between the sums of the fleet totals and the reported total inventories (20 units) in both 1960 and 1965 are not fully explained in the respective editions of *The Military Balance* – although the 1960–61 edition does indicate that eight units were then based in Albania.

The increased importance of submarines in the Northern Fleet is a reflection of two parallel trends. One of these is a decline in the submarine strength of the other three fleets, due largely to the retirement (without replacement) of older short- and medium-range torpedo attack types, primarily the WHISKEY. Between 1960 and 1975 the submarine force of the other three fleets decreased from 270 units (or 63 per cent of the entire Soviet submarine inventory) to 165 units (or 48 per cent). During the same period the Northern Fleet submarine force was increasing. It had 140 units (or 33 per cent of the total) in 1960. Today it has 175 units, or slightly over half of all the submarines in the Soviet Navy.

This numerical increase in Northern Fleet submarine strength is due largely to the introduction of modern ballistic missile-launching units. The retirement of the WHISKEYs in this fleet, however, masks an equally significant qualitative increase in its general purpose component. By 1975 more than half of the entire Soviet inventory of the following modern cruise missile and torpedo attack submarine classes was in the Northern Fleet: CHARLIE-class SSGN, ECHO-class SSGN, JULIET-class SSG, VICTOR-class SSN, NOVEMBER-class SSN and FOXTROT-class SS. And the Northern and Baltic Fleets together had more than 80 per cent of all the Soviet FOXTROTS – their primary diesel-powered torpedo attack class.

This is, of course, no accident. It reflects (1) the existence of combat tasks in and around the North Atlantic that the Soviets feel must be performed, (2) Soviet recognition that these tasks are best performed by submarines perhaps because of the nature of these tasks, perhaps in recognition of the threat posed to Soviet surface combatants and aircraft by NATO land-based air forces and (3) the dictates of geography. Access to the Atlantic from the Baltic can be denied any kind of ship; denying submarines access to the Atlantic from the Barents, on the other hand, is a much more difficult undertaking.

Deployments

This part of the discussion deals with the operations of all Soviet naval forces in the Atlantic, both North and South. The available data do not permit focusing on the activities of just one fleet; they also treat the Atlantic as a whole, including the Caribbean, but not the Barents.

Once again, there are two ways to describe the subject; and, once again, both approaches are employed here. The first is to present a gross characterization of the changing volume of Soviet naval activity in the Atlantic, from 1964 through 1974 (see Figure 1). The unit of measurement is annual shipdays, and these figures aggregate all Soviet naval ships: submarines, surface combatants, auxiliaries, and naval-subordinated merchant ships. For example, 100 shipdays can be generated by one ship operating in the area for 100 days, or (less likely) by 100 ships operating in the area for one day.

Figure 1: Soviet Naval Shipdays in the Atlantic: 1964–1974

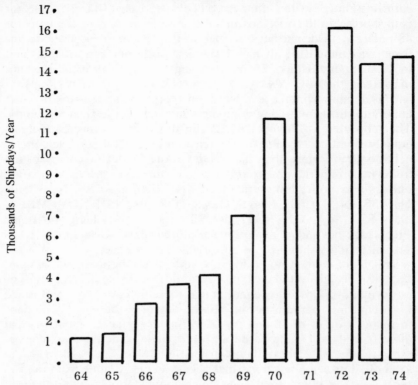

Sources: Robert G. Weinland, 'Soviet Naval Operations – 10 Years of Change,' Arlington, Va: Center for Naval Aanlyses, Professional Paper 125, August 1974; Subsequent Congressional Testimony; and Office of the Chief of Naval Operations, *Understanding Soviet Naval Developments*, Washington, D.C.: Government Printing Office, 1975.

The Atlantic shipday total has grown from 1,200 in 1964 to some 14,600 in 1974, an increase by a factor of 12 or so. For comparison, the world-wide Soviet shipday total has grown by a factor of 15 in the same interval. This process of growth has been anything but constant. Atlantic operations were relatively insignificant in 1964 and 1965, but increased gradually from 1966 through 1968, reflecting Northern and Baltic Fleet participation in the establishment of the Mediterranean Squadron.[3] The primary period of growth in Atlantic operations occurred between 1969 and 1971, reflecting, in order of their occurrence, the initiation of semi-annual replacements of the Northern Fleet submarine contingent in the Mediterranean,[4] YANKEE-class SSBN patrols in the Central Atlantic,[5] and periodic deployment to Cuba,[6] all three in 1969, as well as the establishment of the West Africa patrol in late 1970.[7] Manoeuver *Okean* also made a significant contribution to the 1970 figure.[8] The rate of growth in these operation began to decelerate in 1972 (from a 30 per cent increase between 1970 and 1971 to a five per cent increase between 1971 and 1972). This flattening of the curve probably reflected the achievement of a steady state in YANKEE patrol activity. The figures for 1973 and 1974 show both a decline from the 1971–72 peak period and considerable stability. The decline probably reflects a reduction in the forces operating off West Africa. The stability may indicate that Atlantic operations have attained some degree of maturity.

Gross characterizations like these tend to mask essential information, such as the fact that the level of Soviet naval activity in the Atlantic fluctuates widely. There are two distinct patterns in their operations. There is a 'core' steady-state forward deployment in the Atlantic, in which a small force of combatants and auxiliaries from the Northern and Baltic Fleets is maintained 'on station' more or less continuously. Intermittent deployments of combatants and auxiliaries from these two fleets as well as from the Black Sea Fleet and the Mediterranean Squadron augment this 'core' for specific operations. When those operations are concluded, however, the augmenting forces return to home waters, leaving the 'core' behind. This steady-state 'core' deployment is surprisingly small, given the forces nominally available for operations there. So, for that matter, are most of the intermittent deployments.

One way to illustrate how little the Soviets actually operate out on the high seas is to calculate what the 1974 shipday total would have been if they had deployed and operated the entire combatant forces of the Northern and Baltic Fleets exclusively in the Atlantic that year. The US Navy attempts to keep its strategic offensive forces at sea roughly half of the time, and its general purpose forces at sea roughly one-third of the time. If the Northern Fleet had operated in this manner in 1974, it would have generated 31,500 combatant shipdays at sea. If the Baltic Fleet had done this also, it would have generated an additional 11,330 combatant shipdays at sea. That would have made 42,800 combatant shipdays in all. However, the actual 1974 Atlantic shipday total for all types, combatants and auxiliaries, was only 14,600. Clearly, neither fleet spent very much of its time there. (These calculations should not be taken too seriously. They assume that the US Navy practice can serve as a norm for evaluating Soviet practice. It cannot. They also assume that the Northern and Baltic Fleets can operate in the Atlantic without first transiting the Barents and Baltic. That, of course, is impossible. These shortcomings do not, however, make the illustration any less useful. The 14,600 Atlantic shipday total still represents only one-third of the hypothetically available operational time, and since the Atlantic total includes an unspecified but probably not insignificant number of shipdays spent in the Atlantic by Northern and Baltic Fleet auxiliaries as well as combatant and auxiliary units of the Black Sea Fleet, the point of the illustration is actually stronger than stated above.)

This is an appropriate place to turn to the second approach to characterizing deployments, which is to attempt a 'snapshot' description of the forces operating in the Atlantic on 'any given day.' These will be the 'core' forces of the steady-state forward deployment (see Table 3).

Table 3: Nominal Steady-State Forward Deployment Force*

Strategic Offensive Forces .... 3–5 Units
   2 Y Class (off Bermuda)
   0–1 H (or G) Class (off Newfoundland)
   1–2 D Class (probably in Barents Sea)
General Purpose Forces .... 7–15 Units
   1–3 Torpedo Attack/Cruise missile submarines
   0–2 Surface Combatants
   2–3 Intelligence Collectors
   1–2 Hydrographic/Oceanographic Research ships
   2–3 Space Research ships
   1–2 Rescue/Salvage tugs

Total: 10–20 Units
(Plus units in transit
to/from assignment)

* A composite of information presented in Congressional testimony, official releases, and press reporting (*Daily Telegraph* (London), *New York Times, Washington Post* and wire services).

There will be three to five SLBM submarines 'on station' (counting DELTA-class SSBNs, which if one takes a strict view of what constitutes the Atlantic, have yet to be reported operating there).[9] There will also be one to five general purpose force submarines and surface combatants and six to ten auxiliaries 'on station'. That makes 10 to 20 units 'on station', plus one or more additional units in transit to and from these assignments. The total combatant contingent of four to ten units is really rather small. This is even more apparent when one considers that there are 325 combatants in the Northern and Baltic Fleets to draw on for Atlantic deployments.

If the entire 'core' force in 1974 consisted of 20 units, which is the 'high' end of the nominal range cited, that would account for exactly half of that year's total Atlantic shipdays. The other half of that figure would be accounted for by intermittent deployments. These fluctuate too widely in strength, composition and duration to permit succinct description. It is possible, however, to outline the purposes for which these intermittent deployments are undertaken.

One must begin by differentiating between deployments of Northern and Baltic Fleet units to the North Atlantic and their deployments to other than the North Atlantic, i. e., to the Mediterranean, West Africa and Cuba (all of which are outside the realm of this discussion).[10] In considering deployments to the North Atlantic, one also should distinguish between those undertaken wholly at Soviet initiative and those undertaken in reaction to the operations of NATO forces. The former Soviet initiated deployments undoubtedly represent the majority of their Atlantic activity, and encompass five more or less distinct types of operations: alerts, exercises, workup and training cruises, equipment test and evaluation, and inter-fleet movements.

Soviet Naval Air Force deployments and operations in the Atlantic also deserve brief mention. Once again, there are two distinct modes of operation: 'out-and-back' flights and temporary forward deployments. The former are flights originating and terminating in the Soviet Union and consist largely of BEAR-D open-ocean reconnaissance activity, BEAR-F and MAY ASW missions, and simulated strike operations conducted by BADGERs during exercises. Since 1970, SNAF BEAR-Ds have also made intermittent temporary deployments to Cuba, and since 1972 to Guinea, staging from there for one or more reconnaissance flights in the Central Atlantic before returning home.[11]

Activities of Deployed Forces

Very little information is available on the specific activities undertaken by Soviet forces when deployed. Next to nothing has been published about the strategic offensive forces. Presumably, they simply transit to, within, and from, their patrol areas, which are reportedly now located within missile range of the United States (the patrol areas of the GOLF-class SSBs and HOTEL-class SSBNs that deployed prior to the introduction of the YANKEE were apparently located out of range of the United States).[12]

A bit more has been published about the activities of the general purpose force submarines. They also patrol. They participate in exercises. And they occasionally attempt to trail Western submarines, witness the reports of undersea collisions that have surfaced from time to time.[13]

Quite a bit more information is available on the activities of surface combatants (which are, after all, readily visible). They do not as a rule conduct patrols. They apparently deploy primarily to participate in exercises, and they play an active role in shadowing major NATO forces.

Naval aircraft are also active in locating and conducting reconnaissance of major NATO forces. In addition, they participate actively (and visibly) in exercises.

Very large forces deployed to the Atlantic (and to other ocean areas) for Manoeuver *Okean* (Ocean) in April 1970 and again for Exercise *Vesna* (Spring) in April 1975.[14] However, these were highly unusual operations, worldwide, centrally-controlled operations that, in the light of the publicity given them by the Soviets, appeared to have been undertaken primarily as demonstrations (for both domestic and foreign audiences). Exercise *Sever* (North) in July 1968, although confined to the Barents, Norwegian and North Seas, the Northeast Atlantic and the Baltic, also involved extensive deployments.[15] It, too, gave the appearance of being primarily a demonstration.

In contrast, the remaining eight major fleet exercises conducted in the region over the last eleven years (see Figure 2) involved more modest forces, deploying for relatively short periods, and operating without publicity.[16] With one possible exception (discussed below, p. 70) these other exercises appeared to have more standard formats and objectives: training, and the test and evaluation of weapon systems and tactics.

Figure 2: Major Soviet Naval Exercises in the Norwegian Sea/ North Atlantic: 1965–1975

Year	JAN	FEB	MAR	APR	MAY	JUN	JUL	AUG	SEP	OCT	NOV	DEC	DURATION (Weeks)
1965				█		█	*						2+3
1966								█					1
1967				█	*								2 1/2
1968							█	* (Sever)					2
1969													
1970				█ * (Okean)									2
1971							█						1
1972													
1973				█ *	█ *								1+1
1974					█ *								1 1/2
1975				█ * (Vesna)									1

* = Interfleet Exercise

Source: Based on information released by U.S. Navy.

What these forces do during fleet exercises is fairly well known. It also appears sufficiently stereotyped to permit description of a standard scenario.

First of all, the exercise participants deploy into the Atlantic. The initial phase of actual exercise play appears to take place in the Norwegian Sea and to be anti-submarine warfare (ASW)-oriented. The next phase features coordinated sub-surface, surface, and air strikes against an 'aggressor' surface task force group moving to the Northeast through the Greenland-Iceland-U.K. gap. Exercise *Vesna* (and possibly also Manoeuver *Okean*) included an additional phase of convoy exercises, although whether these operations were oriented primarily toward convoy attack or convoy defense apparently never became clear.[17] The penultimate phase of some exercises (*Sever* and *Okean*, in particular) has featured an amphibious landing on the Rybachi Peninsula and amphibious landings have also been conducted in the Baltic.[18] At the conclusion of exercise play the participating units normally return directly to base.

LIKELY EMPLOYMENT

What we have seen thus far are large and capable forces that are directed primarily toward conducting combat operations in the North Atlantic – but do not as a rule operate there, and have not yet engaged in combat. This obviously complicates the problem of specifying how they are likely to be used in wartime. It also makes it imperative that we address the questions of their likely employment not only in wartime but in peacetime as well. The wartime scenario under discussion is a major war in Europe.

The obvious way to proceed is to start with those aspects about which the most is known (or can be assumed with confidence) and progress from there toward the realm of pure speculation. That implies beginning with the missions and *modus operandi* of the strategic offensive forces in peacetime (about which we can safely make some fundamental assumptions). Next comes the employment of the general purpose forces, first in peacetime (about which most is known) and then in wartime (where their exercise activities provide a basis for extrapolation). Finally, we shall return to the wartime employment of the strategic offensive forces (about which there is no firm indication of what is known).

Strategic Offensive Forces in Peacetime

The peacetime mission of Soviet naval strategic offensive forces appears to be deterrence, to be achieved by the establishment and maintenance of an assured minimum destruction capability *vis-à-vis* the United States. In order to carry out this mission, the Russians maintain a small proportion of their SLBM force deployed within strike range of the United States. They retain the majority of the force in home waters. In all probability, they do this so that, should strategic strikes appear imminent, a larger proportion of the force will be capable of putting to sea and remaining there for a long period.

General Purpose Forces in Peacetime

Soviet naval general purpose forces clearly have two peacetime missions. The first is to *establish and maintain readiness* (both materiel and positional) to contest control of the seas with NATO forces in certain areas and at certain times. Their objective in this is to limit the damage that NATO forces are capable of inflicting on the Soviet Union, its allies and its overseas interests, and to assist other components of the Navy and other branches of the Soviet armed forces in the prosecution of their respective missions. Their second peacetime mission is to act as instruments of Soviet foreign policy in the protection and promotion of Soviet overseas interests.

In order to accomplish these missions, they deploy and operate in response to the presence of major NATO naval forces in their area of responsibility. They also deploy to engage in the training and exercising necessary to establish and maintain combat readiness.

General Purpose Forces in Wartime

The most likely wartime mission of Soviet naval general purpose forces in the North Atlantic theatre is to execute those tasks for which they have established readiness in peacetime. The most important of these are defensive in character. For the Northern Fleet, this would mean conducting strikes against NATO carrier and amphibious task forces approaching and operating in the Norwegian Sea. It would also mean conducting ASW operations in the Barents and Norwegian Seas, to protect the Soviet SLBM force against potential attrition by NATO general purpose forces (and, in all probability, to attrite any NATO SLBM forces encountered).[19] Once these defensive tasks were accomplished, it might also mean mounting an anti-shipping campaign against NATO's sea lines of communication (SLOC) in the Northern and Central Atlantic, if the war were to continue long enough for the SLOC to be established, and if adequate forces continued to be available for such operations. In order to accomplish these tasks, the Northern Fleet would have to secure ready access to the North Atlantic which, at least, would entail denying NATO the use of its facilities located astride the GIUK gap, as well as those in Norway. The Baltic Fleet would be devoted largely to providing direct support to the seaward flanks of the ground forces. It would also have the task of securing control of the Baltic exits.

Strategic Offensive Forces in Wartime

The likely wartime mission of the SLBM force can only be to establish and maintain maximum readiness to launch on order, and to launch as soon as practicable after the order is received. Precisely what this implies, however, is not at all clear. Given its current capabilities and *modus operandi* – a minimum number of units deployed within range of its most likely targets; a maximum number of units retained in home waters – the Soviet SLBM force cannot be the vehicle for a disarming first strike. There is reason to believe, but little unambiguous confirming evidence, that some or all of it might be withheld for purposes of intra- and post-war bargaining. [20]

IMPORTANT QUESTIONS

Among the issues that deserve further consideration are (1) the nature of the wartime requirement for Soviet naval forces to have access to the Central Atlantic, and (2) the extent to which Soviet naval forces are being employed for purposes of political influence in the North Atlantic theatre. Let us address them in that order.

Access to the Central Atlantic

Do the Soviets really have to transit the Greenland–Iceland–U.K. gap in order to accomplish their wartime missions? The answer, given their present forces, missions, and expected opposition is 'yes'. It is mandatory, both before and during conflict, that both the strategic offensive and general purpose forces *are able to* move into strike range of their targets. However, the extent to which they *will* be able to do so once conflict is initiated is not at all clear. Exit from the Baltic can be denied; and the North and Norwegian Seas are likely to be hostile environments in order of decreasing severity for surface ships, aircraft and submarines.

This situation is obvious to the Russians, and the resulting incentive to reduce (if not eliminate) the requirement to transit into the Central Atlantic in wartime has already produced two moves: an attempt to establish forward support facilities (if not bases) for submarines (most likely those of the SLBM force) outside the gap (for example, in Cuba), and the acquisition of submarine-launched ballistic missiles of sufficient range to make transiting the gap to strike their targets unnecessary. Those same incentives have also affected the structure of the general purpose forces based inside the gap – witness the heavy emphasis on submarines in the Northern Fleet. And they have affected the selection of the Navy's primary strike forces: submarines and aircraft. The same incentives may also lead to the eventual acquisition of a reconnaissance satellite-supported, land-based, anti-ship ballistic missile – say, a system that would mate the guidance and control package of the SS N–13 to an intermediate-range booster. This would be employed for strikes against time-urgent, high-value targets such as aircraft carriers.[21]

The introduction of the DELTA/SS–N–8 system has relieved some but not all of this pressure. First, the YANKEE/SS–N–6 system which has slightly over a third of the range still makes up the bulk of the SLBM force, and these units must move through the gap and out into the Atlantic in order to strike targets in the United States. This situation will prevail for several years to come. Second, the DELTA is probably no less vulnerable than is the YANKEE to NATO undersea warfare forces – should the West decide to undertake a damage-limiting campaign against withheld SLBM forces.

In the long run, of course, as the SS–N–8 replaces the SS–N–6 as the principal booster, launch-platform attrition will become less of a problem for the Russians. These new forces, surged from their bases, can be on station sooner; and it will be feasible to disperse them beforehand over a wider area, into the South Atlantic, for example.

One collateral benefit of the acquisition of a DELTA/SS–N–8 force deserves mention. It will make a sustained attack against NATO's Atlantic SLOC a somewhat more attractive proposition. As long as the Soviets have to rely on the YANKEE/SS–N–6 combination, which must transit the ASW barriers they can expect NATO to establish to protect the SLOC, collateral attrition of their deterrent would occur. They cannot afford to sustain such losses in those circumstances. Once they no longer need to transit that barrier, however, attacking the SLOC begins to make some sense assuming, of course, that they feel a compelling need to do so (which they may or may not, depending upon how long they think the war will last) and think they have the forces to do it.

Political Operations
The North Atlantic is generally considered to be a primary theatre of naval strategic offensive and defensive operations; and the Soviet naval

forces that operate there are generally considered to be wholly committed to these tasks. Meanwhile, it is widely recognized that Soviet naval forces operating in the South Atlantic, the Mediterranean, the Indian Ocean, and even the Pacific, have been employed, in varying degrees and at various times, as active instruments of Soviet foreign policy. Why is this not the case in the North Atlantic? Or is it?

It must be granted that over the last twenty years or so there have been fewer opportunities for direct Soviet political involvement in the affairs of the countries on the North Atlantic littoral than, say, those on the Mediterranean littoral. Rigid bipolarity, NATO cohesion and the absence of divisive issues in which naval power was a relevant factor left the Soviets little room for political manoeuver.

However, all three of those factors have been changing in recent years. Rigid bipolarity may not be completely gone, but detente is clearly with us – at least for the present. NATO cohesion is not what it used to be. And divisive issues have arisen in which Soviet naval power already has, or readily could become, a relevant factor – a development that has been recognized explicitly by the Soviets.[22] One of these issues concerns the share of NATO resources required for defense of its Northern flank. Another is the allocation of oceanic and seabed resources.

Re-examination of the recent operations of Soviet naval forces in the North Atlantic, with an eye to both their military *and* their political significance, leads to the suspicion that these forces as well have, in fact, already been used as instruments of international political influence on at least two (and possibly more) occasions. The first of these was in July 1968, via Exercise *Sever*, and appears to have been a part of a larger Soviet effort to deter Western intervention (or counter-intervention) in Czechoslovakia. The second occasion was in May 1973, via another major fleet exercise, and may have represented an attempt, if not to become a player in, then at least to influence the course and outcome of, the 'Cod War' between Iceland and the United Kingdom. *Sever* was one part of a widespread and unusually blatant influence attempt. The May 1973 exercise, on the other hand, represented a limited and much more subtle influence attempt, and serves as a better illustration of the evolving Soviet *modus operandi* in politically oriented operations. It is also more representative of the occasions for direct Soviet political involvement that are likely to arise in the North Atlantic theatre in coming years.

Attempted Soviet involvement in the 'Cod War' is, nevertheless, a topic that must be approached with some delicacy and considerable precision. One wants to avoid 'finding' something that did *not* exist; and equally, one wants to avoid not finding something that *did* exist.

69

What, precisely, is the charge against the Soviets? At the minimum, it is that they attempted to take advantage of the 'Cod War' to further their own interests, their probable proximate objectives being to demonstrate their ability to provide support to Iceland, and to deter the United Kingdom from further escalating its role in the conflict; their likely ultimate objective being to split Iceland from NATO, thus denying NATO further use of critical sea control support facilities located there. What precisely did the Soviets do to merit this accusation? They deployed powerful and highly visible naval forces (reported to consist of 10 cruisers, frigates and support ships and an equal number of submarines) into an active conflict area (Icelandic waters) at a critical juncture in that conflict (immediately after the Royal Navy's role changed from indirect to direct participant as its frigates entered the 50-mile zone) and during the peak period of activity.[23] Figure 3, which is based on press reports of British and Icelandic actions, spotlights that point in the 'war' in terms of monthly totals of cooperative and conflictual actions.

The evidence in support of this charge is purely circumstantial. It consists of three sets of facts: the conflict was intense and had significant potential for further escalation, the Soviets were there, and they

Figure 3: Reported Actions of the United Kingdom and Iceland During the 'Cod War': September 1972–October 1973

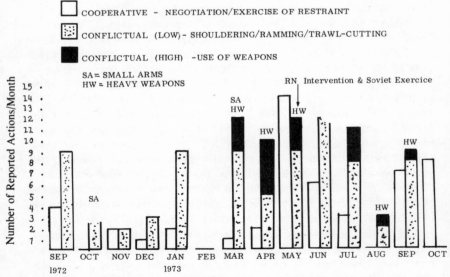

Sources: *New York Times, Washington Post, Daily Telegraph* (London).

did not *need* to be there. First, they had already conducted one major fleet exercise in the North Atlantic-Norwegian Sea area that year, in early April. Holding another one shortly thereafter was not standard practice (see figure 2, above). Second, even if there were legitimate grounds for staging a second exercise, they could have delayed it, or held it in a different location. There is ample evidence of their having done precisely that in previous situations, where to follow through with an exercise might have entailed entanglement in events which they clearly wished to avoid.

Did they succeed in this apparent influence attempt? There is no evidence that either party to the conflict was influenced in any way by their action; and the eventual outcome was certainly at variance from postulated Soviet objectives. The important question is not, however, the degree to which the Russians succeeded in influencing the actions of the parties to the conflict, but the extent to which their actions represented an *attempt* to do so.

IMPLICATIONS

If we assume that there is an essential adherence to the SALT agreements, and there are no fundamental breakthroughs in ASW technology, the Soviets can be expected to do at least some of the following with regard to their naval forces and their use of them in the North Atlantic theatre.

Forces and Missions

The Russians will continue to rely primarily on submarines for the accomplishment of the Navy's strategic offensive and defensive missions. Given this inclination and NATO's air superiority, and its concomittant ability to control at least the sea surface, submarines will continue to be the main component of the Soviet forces allocated to the North Atlantic. But, should the Soviets succeed in developing an effective land-based, long-range anti-ship missile system, that system would be given the primary responsibility for strategic defence. Should this occur, and perhaps even if it does not, the Russians might well reorganize their forces along functional lines, creating: a 'Strategic Offensive Force', encompassing the majority of the resources of the Strategic Rocket Troops, the Navy's SLBM component, and Long Range Aviation; and also a 'Strategic Defensive Strike Force', consisting of the remaining resources of the Strategic Rocket Troops (armed with the anti-ship missile postulated above), some SLBM submarines armed with the SS N–13 or its equivalent, the Soviet Naval Air Force, some of the strike aircraft of the Long Range Aviation, and most of the Navy's general purpose force combatants. The former would retain the present strategic targeting and could be expected to be under the operational control of the General Staff; the latter would be targeted against threatening naval forces and would be under the operational control of the Navy.

Deployments: Given the assumptions noted above, and in the absence of an operational land-based anti-ship missile system, the Russians are likely to make few significant changes in their peacetime North Atlantic deployments and operations (although the DELTAs might well follow the YANKEEs into the Atlantic at some point). Should such changes occur, they probably would be reflections of events outside the theatre, for example, modifications in the *modus operandi* of the Mediterranean or Indian Ocean squadrons, and consist for the most part of increases or decreases in the number of transits to and from those other areas by units of the Northern and Baltic Fleets.

Employment: In peacetime, intensive and extensive use of naval general purpose forces for political influence purposes is a distinct possibility. However, while the political environment has become more amenable, there remain inherent limits to the growth of such activity in this theatre. Obvious Soviet attempts to exploit divisive issues in the Atlantic tend to have a counter-productive impact (for the Soviets) on NATO cohesion.

It is entirely possible that as they gain confidence in the security of their SLBM force, and as both NATO and the Warsaw Pact improve their capabilities to conduct extended conventional warfare on the Central Front, the Russians will place increased emphasis on their capability to interdict NATO shipping across the Atlantic. Whether, once embroiled in such a conflict, they would actually employ their forces in this manner remains problematical, however, and is likely to depend on both what happens to their SLBM force and the extent to which they think attacking the SLOC would materially affect NATO's capability to continue fighting. Significant withdrawals of American ground forces from the continent, or reductions in the stockpiles of materiel prepositioned there, would tend to increase the potential attractiveness to the Russians of such a campaign.

SUMMARY

The strategic offensive forces of the Soviet Northern Fleet, and the general purpose forces of the Northern and Baltic Fleets are large, modern, and fully capable of 'ruining the whole day' for any opponent. There are, however, fundamental limitations on the capabili-

72

ties of these forces, some imposed by the capabilities of their potential opponents, some imposed by geography, and some the result of Soviet decisions.

Examination of the composition of these forces, and their relationship to the total Soviet naval force structure, shows how the Soviets attempt to compensate for these limitations. In the critical areas – the SLBM force, anti-ship missile launchers, and more capable nuclear-powered submarines – roughly two-thirds of the entire Soviet inventory is based in the Northern Fleet. This emphasis on submarines in the Northern Fleet is a reflection of two of those limitations: Northern Fleet units have more direct access to the open seas; but in order to get there, a necessity if they are to accomplish their missions, they must transit the GIUK gap which, given NATO capabilities, would not be a healthy place in wartime for either Soviet surface combatants or aircraft.

Examination of the peacetime operations of these forces in the North Atlantic reveals two distinct patterns in their deployments. A small 'core' of strategic offensive and general purpose forces is maintained continuously on station. From time to time and for specific tasks this 'core' is augmented by additional general purpose forces, which return to home waters once their objectives have been reached. Given the large forces available for employment in the Atlantic, however, both this 'core' and its augmentations tend to be modest in size. This probably reflects a conscious Soviet decision to husband their resources, perhaps for surge deployment when conflict appears imminent.

Further examination of the peacetime operations of these forces reveals that, while their primary mission is to establish and maintain positional and materiel readiness to defend the Soviet Union in various ways should conflict occur, the general purpose component also serves in peacetime as an active instrument of international political influence. One example of this latter usage may be represented by Soviet actions during the 1972–73 'Cod War' between Iceland and Britain, where they appear to have attempted to influence the outcome of the conflict for their own benefit. Given the changes that are taking place in the international environment of the North Atlantic, 'political' operations could well become a regular feature of Soviet naval activity there.

NILS ØRVIK

Canada and North Atlantic Security

While most of the European nations have displayed a growing concern over the recent increase in Soviet military capabilities, particularly in the areas adjacent to the Arctic, these changes in the strategic environment have received little attention in Canada. With the exception of a brief period during World War II, Canadians have tended to see the Arctic as a basically national issue where the major problems are connected with the exploration and exploitation of natural resources; how to get them out of the north, how to assure a reasonable share for the native peoples and how to avoid a conflict with the Americans over the long-term distribution of these vital energy resources.

Experts and groups with special interests are also concerned about such issues as the 200-mile economic zone, pollution, fisheries limits and the implications of increased international interest in the Arctic Ocean, the Beaufort Sea and the Davis Strait for Canada's policy in the years to come. The growing traffic of nuclear submarines under the northern ice cap and the fact that their access and exit routes may bring them close to Canadian shores have raised questions about Canada's ability to control these as well as her responsibility for doing so and what the price would be. At the moment, a major part of the air and sea surveillance is being carried out by the armed forces.[1] These forces are fairly small, but seem adequate in present conditions. Difficult problems would emerge if the demand for surveillance and effective control of the areas should increase to levels where icebreakers, intensive overflights and ground parties might be deemed necessary; should the responsibility rest with civilian authorities, such as the Department of Transport, the Royal Canadian Mounted Police or other government institutions which share the sea and air capabilities in northern waters, or should the armed forces take the full responsibility for doing all parts of the job?

Although there is considerable interest in these matters, most people do not see them as problems of national security in a traditional sense. There is, of course, the general awareness of the great neighbour to the south, the United States, and some uneasiness as to what demands the Americans might make on Canada in a crisis. This anxiety reflects a general concern, applying right across the board from capital investments by American companies, cultural protection of periodical and mass media, energy and industry problems to matters of more direct relevance, such as the famous *Manhattan* voyage in 1969.[2]

The uneasiness about the United States, whose towering presence in the North American continent cannot be ignored anywhere in Canada, is certainly relevant to national security. But it is not seen as a military threat; the idea that the Americans, in a fit of desperation over, say, energy shortages, might invade and occupy Canada belongs to science fiction.[3] It is worth noting that most of the people who are worried about the future implications of the disparity between the two North American countries are also well aware of the close ties that exist between them, their many common interests and the need for close co-operation on many levels.

Thus, the Canadians are probably the only people within the Western defence community who can say in all honesty that they see no direct threat to the nation's security, though this does not exclude a possible future threat. But national security is not seen as an immediate problem. If a threat should ever appear, it will emerge from a sequence of events in Europe or elsewhere; then, in the second or third stage, Canada's security might become an issue.

If one defines the threat as an armed attack on the nation's territory, this popularly held opinion seems basically correct. Only two countries, the Soviet Union and the United States, have the capability to stage an invasion of Canada and neither of them are likely to do so. There are a number of obvious reasons why the Americans would not consider using force, and the rate at which the United States has been losing friends elsewhere in the world since World War II will make her more dependent on those living close by, where ways of life and cultural ties merge with common interest. The Soviet Union is another matter; she already controls more than half of the shores that encircle the Arctic Ocean. With her expansionist leanings, she could be expected to welcome any opportunity to extend her control to parts of the Arctic that belong to Canada. But, although such aspirations may exist at blueprint level, this would not be the appropriate time to put them into effect. The Arctic seems at present the least likely place for a Soviet probe. As we shall see later, one of the Soviet Union's highest priorities at the moment is to avoid any move that might involve the risk of an armed conflict with the United States. Seen from Washington, the Canadian Arctic is a sensitive area. The remoteness and the inacessibility of that region often clouds the fact that it is very much a part of the North American Continent for which the United States feels directly responsible. This brings us right to the core of the Canadian security problem. There is no direct Soviet threat to Canada, because the United States, in the foreseeable future, would not allow the Russians to get anywhere near positions in the area that might endanger the security of the North American continent. Thus, as the situation exists today, the public is right. But in a period of rapid change, a static position is not satisfactory. Rather than make a short-term assessment as a basis for long-term policy, it is best to focus on the indirect threat, which may follow the changes that are taking place today and tomorrow.

The indirect threat might be explored in two major directions. One is the traditional concept of a gradual involvement in Europe through Canada's participation in NATO. The standard hypothesis is a Soviet attack against central Europe, which would engage the Canadian forces there, and bring Canada into a major war more or less along the lines of World Wars I and II; the implications of this central European scenario have been hashed over for close to thirty years and need no further elaboration.

The second possibility, which is much less discussed, might be called the Arctic-North Atlantic dimension, in which the Soviet Union shifts its attention from central Europe to northern Europe. Instead of applying the traditional pressures in the German area, she might try simultaneous military, economic and political pressures to pry northern Europe, the North Atlantic islands and part of the Arctic away from the American sphere of influence, first into a strategic no-man's land and then under indirect Soviet control. If this happened, the whole North Atlantic area from the Davis Strait to Svalbard would become a high tension zone. In that case, one would expect American defence activities to move westward, with a high concentration on areas of Canadian responsibility. It therefore seems necessary for Canadians to try to spell out, and analyse in some detail, the variables that are involved in the indirect threat. This suggests an approach which may differ from the conventional one. While the West Europeans, who are within the range of a direct Soviet threat, pay more attention to force levels and actual deployments, the Canadians will have to be more concerned about basic Soviet intentions and the long-term aspects of Soviet strategy.

Our discussion of the interaction of the Arctic and the North Atlantic will be based on the following assumptions about Soviet policy and intentions:

1. That despite its declared policy of détente, the Soviet Union tries to extend her influence over policy choices in western Europe, not by open military surprise-attack as portrayed in the traditional scenarios, but by creating a situation where the direction of national policies in the respective West European countries conform to major Soviet guidelines or at any rate do not conflict with Soviet basic foreign policy objectives. How far this remote guidance could extend would depend on the special circumstances in each country at any time given;

2. That in order to create such conditions in West Europe, the Soviet Union must demonstrate an overwhelming military preponderance of her own forces, and the United States' inability or unwillingness to respond to it in kind;

3. That the Soviet Union must not in the foreseeable future be involved in any major war, particularly not in wars which might entail the risk of large-scale use of nuclear weapons.

The implications of these assumptions is that, in order to pursue these three objectives simultaneously, the extension of influence over policy direction in western Europe must be done so as not to precipitate an armed conflict with any of the major states in the West, particularly the United States. Therefore, a rational Soviet planner might be expected to single out three major goals: the reduction of American forces in Europe; the discouragement of a full-scale American intervention in Europe; and the convincing of West Europeans that military resistance to a Soviet attack would be futile.

The Soviet planners must be reconciled to the fact that, even after reduction of American forces, smaller units of experts and specially trained personnel will remain in Europe to guard supplies and weapons installations. But the presence of some scattered American military personnel would not matter as long as the United States (and Canada) would have to stage a full-scale military mobilization to bring armed forces and equipment across the Atlantic in order to provide a credible counterweight to Soviet and East European forces. Therefore, the core of the matter is not the American forces that are in Europe now, but the high probability of an immediate and massive reinforcement on the scale of American contributions during the two World Wars. If it should become feasible to achieve a modest reduction in American forces in Europe and lessen the probability of their intervention on a large scale, one should not exclude the possibility that the West Europeans might be driven to bilateral arrangements along patterns that are well known in recent Soviet practice. Though intensified détente measures might prove effective in persuading the United

States to reduce her forces in Europe and in preventing the West Europeans from strengthening their national defence efforts, it is not immediately clear how such measures could affect an American fullscale intervention to prevent an extension of Soviet control in Europe. The most promising would be dissuasion by raising costs, which is just another term for deterrence, and could be done in a number of ways. However, if the assumptions on which this analysis is based hold good, there is one important constraint. While pushing for increased influence in western Europe the Soviet Union will not initiate any move that implies the risk of a major war, which most likely would become nuclear. This puts restraints on the available forms of deterrence and makes it necessary to concentrate on those means that would make an intervention by American forces across the ocean seem prohibitively costly without running the risk of any nuclear exchange. This brings us to the transatlantic sea lanes. Any major intervention involves a number of risks, but one condition must be fulfilled: a high probability that the intervening force will reach its destination relatively intact. If there should be any serious doubt about this, some politicians and political groups might find the price of intervention too high.

The issue of human lives has been particularly important in American estimates of the costs of military operations. The concern for lives, lost in what would seem a futile way before there was even a chance to affect the outcome, would almost certainly become a serious political consideration.

Given the notorious preoccupation with central Europe, which has been characteristic American policy, a gradual extension of control in northern Europe, in order to neutralize the Atlantic Islands, would – from a Soviet point of view – have many attractions. The risk of military confrontations would be smaller than in either central or southern Europe. Apart from small American groups at Keflavik in Iceland and Thule and Sønderstrøm in Greenland, there are no American forces permanently stationed in the relevant area. The Scandinavian members of NATO have consistently banned bases for allied personnel and nuclear weapons from their territories; nor are there any sizeable indigenous forces. The uncertainty about the political future of some north European countries which is due largely to their economic problems, imposes heavy restraints on their efforts to keep up with the growing costs of achieving the goals for their national defence forces, and this situation is not likely to change in the near future.

When taking a hard look at the political, economic, and military changes that have occurred in the area during the seventies, one cannot realistically exclude a situation in which there might be no operational American or other allied military bases and only inadequate indigenous forces in the whole Arctic and sub-Arctic areas, from Greenland in the west to Svalbard and north Norway in the east. Even if a withdrawal of American forces were not followed by an immediate establishment of Soviet bases in these countries, the Soviet naval presence and the change in the political climate in the whole area could undermine the security of western Europe and the United States; it would also have profound security implications for Canada.

The task at hand in Canada is not a scramble to fight off an invasion of Nova Scotia, Newfoundland or Baffin Island, but a cool, hard assessment of the indirect threat and the long-term consequences of the Soviet westward move which is no longer a remote contingency, but a hard fact.

If the United States and other NATO forces are no longer able to guarantee safe and unimpeded use of the transatlantic seaways during an international crisis, two basic questions should be adequately answered: how would this affect Canada's relations with the West Europeans, NATO, the European Community and the United States, and what would be the implications for Canada's policy for national defence?

Canada's membership in the Western Alliance has never been seriously opposed by any representative group within the nation and it is not likely to be questioned in the near future. On the contrary, the government has made some determined efforts to establish closer ties with the European Community as well as with the individual member nations. The persistence of these endeavours and the strength and equality of the argument that supports them indicate a new policy direction rather than an exploratory probe. It seems inconceivable that this European orientation would continue to meet a favourable response if Canada could afford to risk being cut off from a seaway connection with western Europe during a major crisis; on the contrary, Canada would then find herself alone with the United States in attempting to safeguard the security of the North American continent. This could pose some problems which might be illuminated by the 'defence-against-help' model.[4]

If the United States should become seriously concerned about the safety of the sea lanes as well as the nation's security, she would have to see the whole continent as one defence area. In these terms Canada is placed between a northern and a southern part of the United States. Obviously, there would be bilateral consultations and discussions of joint problems, including such issues as assessment of the threat and the measures which the new situation would require. If the Canadian government should appear unwilling or unable to meet such requirements, as defined by the United States, the Americans might, in the interests of joint North American security (for which they feel ultimately responsible), take actions which might be thought to infringe upon Canada's sovereignty and lead to a general American influence in matters that are now exclusively Canadian.

There is no indication that the technology relevant to the transatlantic sea lanes has reduced the need for bases. One could, albeit reluctantly, envisage a situation in which most American surveillance and other activities which now take place in the North Atlantic Islands would be moved to various base sites in Canada, ranging from the Maritimes over Newfoundland and Labrador to the Northwest Territories. With the harsher international climate that would seem bound to follow such a deterioration of Western positions, it is probable that American efforts to maintain some freedom of movement in the Atlantic area would have to be greatly increased. This might lead to a much heavier American engagement on Canadian territory than ever before.

This picture may look somewhat unfamiliar today, but it deserves further consideration. The Americans are not likely to be ushered out of the North Atlantic area by a few setbacks in western Europe. Nor is there any certainty that, if asked to leave their bases in the North Atlantic, they would actually do so. They might decide to stay and challenge a Soviet Union posing in a self-appointed role as upholder of the rights of small nations and national minorities in that area. In such a case, Canada could hardly avoid sharing the responsibility for the joint tasks in the North Atlantic. If she had the capabilities available to carry out the tasks that would fall to her, the United States would prefer her to do so. On the other hand, if Canada were unable to carry her share, the United States, in the joint interest of safeguarding the security of the continent, would have to step in and make sure that the job was done – if necessary on American initiative and with American forces. One does not have to dig deep in the tangle of problems involved to see what complications and dangers a 'defence-against-help' situation might imply for Canada. Thus, from a Canadian point of view, policies of national security should aim at avoiding a bilateral engagement to safeguard the transatlantic sea lanes and the security of the North American continent. If one accepts the problem as being relevant, there are no easy and obvious answers.

A CANADIAN FORWARD DEFENCE POLICY?

A serious attempt to avoid the possible complications of a future 'defence-against-help' situation would provide Canada with some hard choices. An active national defence policy that would make a real improvement in the present situation could not be achieved by minor adjustments and rearrangements. It would require a much larger portion of the national budget, and in a period which many see as an economic crisis, with a declared government policy of restraints on all spending, any move toward higher allocations for national defence would cause great political problems and present a tremendous challenge to the political leadership. Therefore, in terms of the so-called political realities, it might seem futile even to discuss it. However, as it is the task of the strategic analysts to make assessments on strategy, rather than on parliamentary constellations, some suggestions will be offered.

In moving into a more active defence posture in the North Atlantic area, some points seem more essential than others. The first, which might seem obvious to most people, is that whatever policy might be initiated on the Canadian side, it will have to be carried out with the United States; close co-operation with her is a clear condition for any matter that concerns Canada's national security. That does not mean subservience and meek submission to any proposal coming from Washington – the number of mistakes and misjudgements that have marked recent American foreign policy indicate her fallibility. But criticism should be constructive and made in an atmosphere of active involvement, with a willingness to discuss and improve. There is so much common ground and so much at stake that it is clearly in the interest of both North American countries to avoid a situation which would require the territorial defence of the North American continent.

The formation of Canadian defence policy will, so far as can be foreseen, take place in the triangular relationship with western Europe and the United States. This is not a new proposition to Canadians, nor does it have to be an unpleasant or hopeless one. It should be remembered that it was a Canadian government which provided the first blueprints for what became the North Atlantic Treaty Organization. However, instead of engaging herself in the immediate problems of protecting the North Atlantic, Canada was projected into the international section of the Western defence community, leaving the protection of her Atlantic 'front yard' to the Americans. The question now is whether this is the time to change the priorities and participate more actively in the immediate defence of the North American continent.

The safeguarding of the transatlantic sea lanes is a political as well as a military condition for the continued existence of NATO as a viable defence arrangement. What has kept it going is the conviction shared by most West Europeans that the North Americans are willing and able to come to their rescue if the Soviet Union should decide to attempt to bring western Europe under control. If there should be any doubt about the ability of the North Americans to keep the sea lanes open, there would immediately be questions about their willingness, which might frighten some shaky West European governments into exploring special arrangements with the Warsaw Pact Powers.

As the Soviet emphasis now seems to be on northern Europe and the North Atlantic islands, there is a need to intensify NATO efforts in that area. A fresh initiative would demonstrate to the Soviet Union as well as to the Western allies the high priority this area enjoys within the Western defence community. Rather than leaving the North Atlantic islands and the adjoining sea passages to the Americans and the British, which has been the case so far, it would be in Canada's best interests to become politically and militarily engaged in an active, forward defence of the Arctic and North Atlantic sector of the Western defence community.

The point which will make our politicians despair is that the intensified effort will have to be made in addition to keeping a viable Canadian force in central Europe. The military value of this unit as it exists today has been questioned, but its political importance is of the very first order. The west Europeans are firmly convinced that Canada's military contributions are vastly disproportionate to her economic status. Whether or not this is the case is almost irrelevant as long as the Canadian government is unable to convince them that this is not true. If Canada wants closer ties with the European Community, it seems necessary to improve her military image within NATO by giving a higher priority to national defence.

THE ARCTIC COMMAND

While the problems related to the Canadian forces in central Europe are more a matter of money than mission, there seem to be no clear conceptions of how Canada might make a larger and more direct contribution to the protection of the sea lanes and the adjacent areas in the North Atlantic region. Canadian naval forces are already heavily involved in anti-submarine activities in the Atlantic. There is also the Canadian mobile force earmarked for the northern flank. What else could Canada do?

One task which is being seriously considered is a reinforcement of Canada's coastal defences for surveillance of fisheries and pollution zone limits and the proposed 200-mile economic zone. Oil rigs and other components of the new development of off-shore resources also need protection. With Canada's enormous eastern coastline, ranging from Nova Scotia to the High Arctic, this will be an expensive proposition. The equipment alone will be very costly, but these problems are common to most coastal states on both sides of the Atlantic. It seems quite clear that unless there should be some sudden and substantial improvement in the world's economy and a major change in the average politician's attitude to defence, it will not be possible to cover the needs for protection of off-shore resources, fisheries limits, etc., and at the same time meet the demands for effectively safeguarding the transatlantic sea lanes. The latter is now one of the main tasks for NATO and, as such, a joint responsibility. On the European side it is now evident that the former task of surveying and protecting off-shore installations is beyond the capacity of any of the coastal states, including Britain. If one aims for credibility, without which the task would have no point, one will have to find some way by which the national efforts can be combined.

As the two tasks of protecting sea lanes and coastal installations are not only related, but inseparable, it seems possible to design some arrangement by which the policies guiding these activities could be co-ordinated. Owing to the tensions and uncertainties, which are still a predominant trait in, for instance, issues concerned with fisheries, a co-ordinating arrangement cannot be created overnight. But when there has been a general acceptance of the principle of the 200-mile economic zone and the individual states have reached some agreement on the fisheries in question, these issues might themselves be the basis for co-operation. With this in mind, it might be worth looking at an arrangement which is tentatively called 'NATO's Arctic Command', whereby the nations who share the responsibility for the North Atlantic area would combine their efforts for national as well as regional security.

Intensifying the effort will increase the costs. But considering the risks of continued drift, the investment might quickly pay off. There are a number of reasons why a NATO Arctic Command,[5] situated in the islands north of Scotland, could serve a useful purpose. First, it would bring the area more into the mainstream of the international discussion of Western security. The lack of planning and preparation in the North Atlantic area is, to a large extent, due to lack of public attention. Central and southern Europe have over the past decades been 'oversold' by NATO as a public relations issue and have received disproportionate attention. This is fully understandable in terms of GNP, populations and political weight and may therefore have been justified in the days when the north was seen as a distant ice barrier on the periphery, posing no pressing problems for Western security. As the strategic focus has changed, it would seem appropiate to adjust the organizational structures accordingly.

83

Secondly, an Arctic Command would be a constructive contribution to the solution of the complex problems which are connected with oil discoveries, as well as the unrest caused by disputes and frictions over fishing rights, extension of territorial limits and various categories of 'zones'. These new problems are now intertwined with the traditional military ones. An Arctic Command placed somewhere in the Shetland Islands would serve a political as well as a functional purpose in focusing attention on the new aspects of North Atlantic defence. The most important aspect of such an arrangement would be the political and the psychological effects, which might work in several directions. The establishment of an Arctic Command in that crucial area would provide a warning to the Soviet Union that the North Americans and the West Europeans are alert to the challenge and that they are not drifting apart; that, on the contrary, they are increasingly aware of the dangers involved in the changes taking place in the North Atlantic, and are adjusting to these new facts with appropriate measures for their common defence. It would also have a reassuring effect on the governments and the people in that area, who have seen the changing trends in the north and have despaired over the fact that their friends did not seem to understand their problems. A physical demonstration of these new priorities might stimulate their willingness to increase their national defence efforts and contribute more to joint defence measures.

Lastly, an Arctic Command with a substantial Canadian involvement could have a positive effect on the general Canadian attitude to matters of national security. Having the strategic importance of 'the European connection' and the transatlantic air and sea lanes spelt out might make national defence and Canada's Atlantic responsibilities a more significant issue than they are today. The practical implications of creating a new command base, whatever it is called and however it is composed, will inevitably produce the old cry of 'Don't rock the boat'. But the boat has been rocking for a long time, and to an increasing extent in directions that do not serve the purposes of the Western defence community. A review of the present structure along the lines suggested above might counteract the emerging image of NATO as a stalemated institution and provide proof of its ability to adapt constructively to a changing situation.

More important than any structural change is a broader view of the problems related to national and international security. Gone are the times when military, political, economic and social problems could be tucked away in neat compartments, and dealt with separately; the problems of national security and defence concern us all and affect every aspect of our lives. What is needed in the present situation is an integrated approach to strategic analysis and operation, in which the relevant military and non-military problems are dealt with in the same overall perspective. This will not be easy but the security of the North Atlantic depends on its success.

ERLING BJØL

The Arctic in Danish Perspective

From the Danish point of view the inter-relationship between the Arctic and the North Atlantic primarily concerns Greenland, then the Faroe Islands, and Denmark herself only indirectly. I intend to treat the military, economic, legal and political aspects in that order.

MILITARY ASPECTS

The military aspects fall into two more or less separate categories; those which are related to the maintenance of Danish sovereignty and those which are related to the global strategic situation. Since the latter are the more direct concern of this book and since the former are closely related to the economic and legal apects, I shall first deal with the implications of the changes in the global strategic situation.

Roughly speaking, I think, one may divide the strategic role of Greenland and the Faroe Islands into four phases. The first one covers the years 1939–45. The second emerges gradually after 1945, with the growing tension between the United States and the Soviet Union, and the increasing American reliance on air power. It fades out towards the end of the 1950s with the perceived incipient Soviet ICBM capability and the increasing American reliance on missiles instead of manned bombers. The third phase, mainly in the 1960s, is the age of the ICBMs. The fourth phase begins towards the end of the 1960s with the age of the SSBNs. But, obviously, there is an overlapping between the various phases, particularly between phases three and four, and it is arguable that a fifth phase has already begun with the appearance after 1972 of the Soviet SS–N–8 missiles.

During most of phase one Denmark was occupied by Germany and the Faroe Islands by the British, whereas Greenland remained under free Danish administration, directed by the independent Washington Legation.[1] Under the Greenland Agreement of April 9, 1941, Ambassador Henrik Kaufmann opened Greenland to the United States for the construction of two air bases, Blue West I at Narssarssuaq near the southern tip of Greenland and Blue West VIII at Søndre

86

Strømfjord on the west coast, in the region of Holsteinsborg, and one naval base at Grønnedal close to Ivigtut. Greenland's main role during this period was as a stepping stone between the North American continent and the British Isles, allowing the transfer of short-range planes by air, which thus evaded the German submarines; although this role ceased with Germany's defeat, the United States nevertheless hung on to the bases. Legally this was possible since the duration clause of the treaty stipulated that it should remain in force 'until it is agreed that the present dangers to the peace and security of the American continent have passed'. The interpretation of this clause might have caused problems, since it might have been argued that, even if the Americans still felt that there were dangers to the peace and security of their continent, the Danes could have replied that these dangers were at any rate not the ones which had been 'present' when the treaty was signed. The issue was, however, never raised, since Denmark joined The North Atlantic Treaty as one of the founding members.

Since then Greenland's strategic significance had changed. The considerable results achieved by American strategic bombing of Germany beginning in the summer of 1944, the atomic bomb, the election of a Republican Congress in 1946, all contributed to making the Strategic Air Command (SAC) the basis of American strategy. Greenland is on the shortest line between the industrial heart of the United States and the industrial heart of the Soviet Union, which was the reason for the construction of Thule Air base, further to the north and closer to the Soviet Union than either Søndre Strømfjord or Narssarssuaq (Thule is almost exactly half way between New York and Moscow). The new base was built according to a new Greenland treaty concluded on April 27, 1951. The Americans built Station Nord even closer to the Soviet Union in Kronprins Christian Land in northeast Greenland. Station Nord is essentially an air strip for the emergency use of bombers returning from the Soviet Union. With the development of longer range bombers and missiles, Thule gradually became less important as an air base, but it acquired a new significance as a warning station in the new ballistic missile early warning (BMEW) system which NORAD completed in 1964. With Clear in Alaska and Fylingdales Moor in Great Britain, Thule became one of the three pillars of the BMEWS. At the same time the distant early warning (DEW) line against manned bombers had its Greenland component with the *Dye* 1, 2, 3 and 4 strung across the inland icecap from Holsteinsborg to Angmagssalik and serviced from Søndre Strømfjord, which is now also an important civilian airport just like Narssarssuaq. Grønnedal naval base was transferred to the Royal Danish Navy many years ago and is now the headquarters of its Greenland Command.

It has been argued that with the appearance of the over-the-horizon back scatter (OTHB) and the early warning satellites in synchronous orbits, Greenland's importance in the North American warning system has been reduced. General Lucius D. Clay, head of NORAD, wrote as late as 1974: 'In the past six years, BMEWS has been joined by a forward scatter over-the-horizon radar missile detection system. The name is confusing but the job it does is good. The tail fire of any missile rising from the Eurasian land mass is detected immediately after launch and reported to our Cheyenne Mountain computers. An eight-site submarine-launched ballistic warning system is also in place on our Pacific and Atlantic coasts. We have early warning satellites in synchronous orbits. They give us instant warning of sub-launched missiles as well as land launches. These four systems, working together, give us greater reliability and capability to meet a diversified threat . . . We must rely on redundancy in our systems. The more different ways you can detect something as speedy as ballistic missiles, the less chance there is of either missing it or mistaking something else for a missile . . . when it may be a phenomenon of nature.'[2]

Considering the submarine threat in the main sea lanes between the Soviet base of Murmansk and the North Atlantic which pass between Greenland, Iceland, the Faroe Islands and the Shetland Islands – the famous GIUK gap – it may be assumed that both east Greenland and the Faroe Islands are of some value to the various ASW systems.[3] But very little information seems to be declassified on this subject, and anyway the appearance of the SS–N–8 type missiles in the Soviet navy changes the dimensions of the threat once more. As the Danes see it, the new threat may somewhat alter the strategic significance of eastern Greenland. Since the SS–N–8s can, for second-strike purposes, be fired from the Barents Sea, their need to pass the GIUK gap diminishes. On the other hand, there might be an advantage in their taking cover under the ice in the Greenland Sea. This would again increase the importance of northern Greenland, since the narrow strait between Greenland and Ellesmere Island is the shortest way from the North American continent to the Polar Sea, and therefore a convenient passage for hunter submarines. The Canadians' frantic efforts to expand their Alert base on Ellesmere Island (which, incidentally, has again made Thule, which is conveniently situated close to a good harbour, useful as a supply base and air field) indicates that their views are similar on this point.

Denmark is not equipped to participate in the technologically extremely demanding ASW. She has, however, felt that the new significance of the Polar Sea demands a more active presence in Greenland's uninhabited North-eastern Area. For years, this part of the island has been policed by the Sirius dog sledge patrols, which during World War II dismantled weather stations that the Germans had stealthily put up on the east coast. Station Nord was handed over to the Danes when it had served its useful purpose for the Americans. Since its original purpose was to keep an emergency landing strip open all winter, it was designed for a considerable crew, but as the Danes have very little in the way of strategic bombers, they did not need a base of that size and found it too expensive to keep open. However, since unknown strangers have been taking an interest lately in Station Nord, it was decided that it would be redesigned for a much smaller staff and re-opened, partly to serve as a supply base for the sledge patrols, partly to improve weather forecasts for the region. Later, the sledge patrols might be assisted by short take-off and landing planes (STOL) of the Canadian type. It was not felt necessary to keep the air strip open all year round. The team which has now been left at Station Nord is equipped to be cut off from the rest of the world for eight months except for radio contact. Furthermore, outposts and depots are spread over the coastal area of the whole of North-east Greenland from Daneborg over Station Nord to Thule, so that it can be policed by the sledge patrols.

The importance attached to north-east Greenland these days can be seen in the fact that the Danish Defence Chief spent some time at Station Nord last summer, and was visited there by Defence Minister Orla Møller, accompanied by his leading staff members.

ECONOMIC ASPECTS
It should be stressed that the upgrading of the Danish military presence in the area is not only related to new developments in global nuclear strategy, even though the Russians have shown enough interest in weather reports from east Greenland to imply that they would not mind having their own weather stations there (a suggestion which has of course been politely turned down by the Danes). Economic and legal considerations are also part of the reason for the increased Danish activity in this desolate part of the world. The increase in oil prices has attracted the attention of the international oil companies to Greenland, but so far concessions have only been handed out for drilling off the west coast between Holsteinsborg and Gothaab, a reasonably ice-free part on the Davis Strait. Some experts see promising geology in the extreme north also, and point out that actually this area is more accessible from the American east coast than is Alaska.

Moreover, the 1972–73 boom in minerals has stimulated some people's belief in the viability of further mining in Greenland, apart from the already profitable exploitation of some lead and zinc deposits at Mamorilik on the west coast; chrome, in particular, could become valuable to the Western Alliance.

LEGAL ASPECTS

In any case, the imminent expansion of coastal states' rights to the seas by, or without, new conventions will create new problems for the Danish defence forces. The extension of the territorial sea to 12 miles will increase the surveillance tasks around the Faroe Islands and Greenland. But this is nothing compared to the tasks which will be created by the establishment of 200-mile economic zones. Again this will concern both the Faroe Islands and Greenland, but not on the same scale or in the same sense. The Faroese, being long distance fishermen, will not be enthusiastic about the Greenlanders' expanding their fishing rights to a 200-mile zone. And as far as scale is concerned, it may suffice to mention that Greenland is an island covering 840,000 square miles, which is nine times the size of Britain and gives an idea of the immensity of the inspection problem should it be surrounded by a 200-mile economic zone (except of course on the west coast and on the part of the east coast where the sea is frozen even in summer).

A particular problem could be created by the extension of the territorial sea to 12 miles in the Kennedy and Robertson Channels, which are at several points less than 24 miles broad. Since, as has been pointed out, these waters might become more and more attractive to submarines, the new Law of the Sea decisions concerning straits will be of great interest. Presumably, neither of the two leading naval powers would like to see the regime of the Danish Baltic straits extended to Greenland.

PRACTICAL ASPECTS

Fishing has so far been the main industry in Greenland, and the G–60 development plan, which was launched at considerable cost in the beginning of the 1960s, was mainly based on expansion of the fishing industry. Climatic change has, however, hit this plan severely; the cod catch sank from 33,000 tons in 1966 to 16,500 tons in 1972, leaving only the shrimp fishing off Holsteinsborg and some salmon as a sizable source of revenue, though not nearly enough to solve the problem of the economic viability of Greenland, which has been a considerable problem for the Danes for years.[4]

90

Even Professor Mogens Boserup, the father of G–60 and a close collaborator of Gunnar Myrdal, a few years ago suggested that the most sensible economic policy for Greenland might be deliberate encouragement of emigration to Denmark instead of pouring $2,000 per Greenlander into the island yearly.[5] This is an argument which has been taken up with some political force by the Danish Poujadist movement of Mr Mogens Glistrup whose party, to everybody's surprise, came into the Folketing in strength in the 1973 parliamentary election. But for the following three reasons this is unlikely to become the policy of the Danish government:

1. Denmark has put so much money into social investments in Greenland that at least the social conditions for a viable modern society can now be said to have been created.
2. The prospects of oil have created a certain optimism. Perhaps oil could become the basis of a viable economy on the west coast, centered around Holsteinsborg. After all, there are only about 50,000 people living in Greenland, and one achievement of the immense social investments is the decline of the birth rate, which makes the growth of the Greenland population manageable.
3. The Greenlanders, particularly the young ones, more of whom are now receiving higher education in Denmark, are developing some kind of national consciousness which, wherever else it may lead, certainly does not improve the possibilities of a depopulation through emigration policy.

The Danes are trying to defuse the new Greenland nationalism, which is also to some extent a new left movement, through concessions. A home rule constitution for Greenland is being prepared, which will increase the authority of the local Diet. But the vision of oil riches, and the fact that even quite tiny territories have in recent years obtained the status of independent states, could create a movement for independence. Such a movement might conceivably be exploited by Warsaw Pact countries for the purpose of infiltration. It seems that the core of the radical Greenland nationalism is to be found among the young Greenlanders who are studying in Denmark and are exposed to the influence of the Danish radical Left. Will they be immune to the subversive activities being spread among young Danes from the Rostock Communist party training centre in the German Democratic Republic?

The present generation of young Greenlanders could prove an explosive element; they are numerous, having been born in the years that followed the Danish crash programme for improvement of living conditions in Greenland, which followed Prime Minister Hans Hedtoft's 'social shock' when he visited Greenland in 1948. This programme brought the high death rate down, whereas the birth rate stayed at the high level of a developing country for several years. This generation of young Greenlanders is furthermore the first one to have had the benefit of higher education, though, unfortunately, mainly in Denmark, as there is only one establishment of higher education in Greenland, a teachers' college in Godthåb. For many of them, the years in Denmark mean a cultural shock of alienation which will often lead them into nationalism and perhaps even separatism. For those staying in Greenland, the economic opportunities are bleak; inadequate training excludes them from most of the better jobs which go to the Danes. It would not be difficult to persuade them that the immense economic difficulties which, as convincingly shown by Dr Christian Vibe, are due to cyclic shifts of climate, are really the fault of the Danes.[6] Unfortunately, Greenland is at present in the meagre years of a cycle, with both fishing and hunting scarce. Should oil and mining prove to be a new basis for the economy, the political atmosphere will not necessarily be improved since again Danes will undoubtedly, if unwisely, be often preferred to Greenlanders, whose traditional culture has not prepared them for the standards of efficiency and accuracy demanded by a modern economy, but rather for survival under the extreme rigours of Arctic conditions. One may perhaps get an idea of the problems of adaptation confronting a people without abstract notions, or even figures, in its language, when one hears that the number nine is translated as *ku-li-ngi-lu-at i-ma-lu-nit ku-lai-lu-at*! On the other hand, indifference to time undoubtedly may be an excellent quality for survival in a climate where winds and visibility may change completely in a matter of minutes, playing havoc with all travel schedules.

It should of course be obvious that a country the size of Greenland with a population of 50,000 could not long exist as an independent political unit without suffering serious encroachment from stronger states if the Danes were to leave. But all the Greenlanders do not necessarily see it that way; a recent poll in Greenland on the flag question showed 90 per cent in favour of the Danish flag, Dannebrog.

CONCLUSION

For strategic, economic and legal reasons – and perhaps for political ones as well – the world is probably going to pay increasing attention to Greenland – without even mentioning the United Nations, which after the dissolution of the Portuguese Empire might soon find itself short of colonial peoples to liberate. These new dimensions of the Arctic problems might call for a closer co-operation among nations such as Canada, Norway and Denmark, which after all have much stronger common than conflicting interests in the area. Such co-operation would be particularly advantageous between Canada and Denmark on the west and north coast of Greenland, but might even be useful between Norway and Denmark, and perhaps Iceland, on the far less hospitable east coast and in its adjacent waters. In the west the Danes and Greenlanders could profit greatly from Canada's Arctic technological know-how. The same might even be true on the east coast in the exploitation of mineral resources and water power. But the Danes would not necessarily only be on the receiving end, especially in techniques of survival under Arctic conditions; the Danes and Greenlanders have considerable experience in adaptation to Arctic conditions, whereas the North Americans have been more inclined to seek protection against them, at some cost to themselves. But both the Canadians and the Norwegians have probably experienced a certain inclination on the part of the Danes to consider Greenland very much a *chasse gardée*; this may, however, be about to change. Greenland's new place in the world is also awakening greater interest in Denmark in the problems of the huge Arctic island, whose fate has hitherto been very much the concern of rather limited circles. Perhaps it will be realized that the whole administrative structure of Greenland affairs is becoming more and more anachronistic.

FINN SOLLIE

The Interaction of the Arctic and the North Atlantic: A Norwegian Perspective

The combination of developments in international law, in offshore technology and in energy prices that led to the oil development in the North Sea, has attracted fresh attention to the possibilities there may be for exploiting resources in Arctic waters. This has further strengthened the general Arctic boom that is underway throughout the northern periphery – from Alaska to the Canadian archipelago and Greenland; from Svalbard and the Barents Sea through the vast stretches of the Soviet North, which alone accounts for almost half the entire Arctic.

Development of new resources on a massive scale will have attendant problems in any region; when the development occurs in a new territory such as an offshore region, problems are multiplied because this is a new kind of development taking place in a region that has not before been a location for permanent operations in fixed installations. When such operations are in the offing in the distant stretches of the Arctic, additional problems emerge, particularly in regard to the environment and in connection with the protection and safeguarding of installations and operations in areas that are so far removed from established centres of industry and population.

The problems of the North are a concern primarily of the Arctic powers themselves, i. e. the USA with Alaska, Canada, and Denmark with Greenland, Norway including Svalbard, and the Soviet Union. However, vital resources in great quantities may be involved that can supply other nations as well – primary among them Europe's industrial states – and this will cause keen interest on their part as well in the Arctic development. Furthermore, to the extent that development in the Arctic may involve strategic problems and affect existing defence commitments, partners to the military alliances will be directly or indirectly concerned.

In the interaction of developments in the Arctic and the North Atlantic, military and strategic implications cannot be separated completely from other effects of the exploitation of offshore resources. In

94

peacetime, policing, control and protection of installations may be regarded entirely as a national responsibility of the coastal state, but in wartime, defence of installations – to prevent their destruction or to insure that they do not fall under enemy control – is essentially an alliance responsibility. A state of readiness must, of course, be established before a crisis develops. There is also the further complication that new developments in weapons systems – such as the Soviet DELTA II strategic submarine equipped with long range missiles (SSN–8) – are now causing a change in the strategic role of such Arctic waters as the Barents Sea. In principle this is a separate development unrelated to the new role of the Barents Sea as a potential oil development area. Nevertheless, the fact that these two separate developments coincide; that new military and new economic and resource interests are developing simultaneously in the same region, makes it difficult to maintain a perfect distinction between the strategic and the economic aspects of the situation.

To add even more to the complexity, the Barents Sea presents a legal problem over and beyond the general delineation of national and international rights in a sea area: should the special rights accorded to foreign nationals in Svalbard, under the Svalbard Treaty, also apply to the continental shelf around the islands? While in a strict sense this is a legal issue, attitudes will be affected by economic considerations and by evaluation of potential strategic implications. Moreover, it is hardly possible to extend to the continental shelf those provisions of the Svalbard Treaty which provide equal economic rights to foreign nationals unless other substantive provisions of the Treaty are to apply equally. Under the terms of the Treaty, Svalbard is a demilitarized zone and, in the words of Article 9, 'it may never be used for warlike purposes'. If that were to apply to a vast continental shelf area around the islands and to the waters above that shelf, the strategic implications – and even more the enforcement obligations – would be quite momentous.

GEOGRAPHIC DETERMINANTS

In looking at these problems from the Norwegian point of view, it is necessary to be aware that the Norwegian perspective is framed by a set of geographic determinants. The most characteristic single feature of Norway's geography is her uncommonly long coastline and the narrowness of her land inside. With a mainland area only slightly larger than that of New Mexico (125,000 sq. miles compared with 121,000) or Poland (120,000), Norway has a coastline as long as that of Western Europe from Hamburg to Cape St. Vincent at the southern tip of Portugal. Not counting the fjords, which would give Norway a shoreline more than 20,000 km. long – or half the distance around equator – the Norwegian coast is 2,650 km., or 1,650 miles long and spans more than 13 degrees of latitude (from 57° 57' 31" to 71° 11' 8"). Including Svalbard, Norway reaches 81° N, and spans 23° of latitude. Remembering that the polar circle is at 64° 33' or less than one-third of the way from the southernmost point to the extreme north, this indicates how much of a northern state Norway really is, a point which becomes particularly clear when we also take into account the seabed areas that are subject to Norwegian control.

Under the terms of the 1958 Continental Shelf Convention, as a coastal state Norway 'exercises over the continental shelf sovereign rights for the purpose of exploring it and exploiting its natural resources.' If we use the shelf edge as a limit – in the case of Norway this will coincide with a 500 to 600 m depth limit – the seabed area subject to Norwegian control is approximately 1 mill. sq. km. If we use the 200 mile exclusive economic zone concept as proposed at the Law of the Sea Conference, the total marine area subject to Norwegian control adds up to more than 2 mill. sq. km. Of this area some 350,000 sq. km. is made up of the zone around Jan Mayen (north of Iceland) and the combined zone around the mainland and Svalbard has an area of 1,720,000 sq. km.

Two important facts are enclosed in these figures. First of all, the Norwegian seabed and sea area is many times larger than the national land territory and secondly, the bulk of this marine domain lies in the Arctic. The Norwegian mainland is almost 324,000 sq. km. and Svalbard slightly more than 62,000, which makes for a total land territory of some 386,000 sq. km. In other words, with a 200-mile economic zone the sea to land ratio is more than 5-to-1. If we consider only the area above the continental shelf plateau, the ratio still is 2.5-to-1. Of this vast marine area, the Norwegian part of the North Sea, which is limited by median lines, has an area of some 140,000 sq. km., while the area to the 200-mile limit between 62° N and 70° N is some 380,000 sq. km. In other words, a full 1,200,000 sq. km. of the Norwegian marine domain lie to the north of Norway. This Arctic part of the Norwegian sea/seabed zone is equal in size to the whole economic zone of Argentina and it is 25 per cent larger than the United Kingdom zone.

Current oil policy and production plans in Norway are based on the resources of the North Sea and call for a maximum production of some 90 mill. tons of oil and oil equivalents per year, or 1,8 mill. barrels a day. At this production rate, the reserves will probably last for 50–100 years, possibly longer. Consumer nations find this production rate

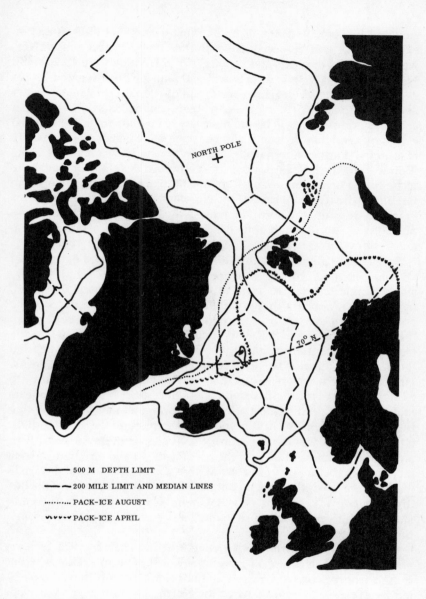

───── 500 M DEPTH LIMIT

─ ─ ─ 200 MILE LIMIT AND MEDIAN LINES

············· PACK-ICE AUGUST

ѵѵѵѵѵ PACK-ICE APRIL

low and wish for more intensive exploitation to help meet their energy needs. By Norwegian standards, however, 90 mill. tons of oil per year is a great deal and involves a major shift in the economy. 90 per cent of this oil will be exported; even if we calculate with an average price of only $8–10 per barrel, the annual export value of this oil (and gas) will be some 45,000 mill N. kr. This means that the value of oil exports alone will be nearly equal to the combined value of all Norwegian exports of goods and services in 1973, which were 48,700 mill. N. kr. By 1980–82 oil exports will account for an estimated 40 per cent of the total Norwegian exports, and of course, oil related industry and offshore activities will make a significant addition. In future decades, therefore, oil will be the main single component in the national economy, and that economy will be based increasingly upon offshore resources. Furthermore, while Norway's new oil economy is being founded on North Sea resources, its continued development will have to be based on exploitation of the northern areas. In other words, in the future the resource base of the Norwegian economy will be shifted gradually towards the north, increasing the role and importance of the northern or Arctic shelf and thereby increasing the nation's dependence upon those regions.

So far, only preliminary investigation has been carried out north of 62° and the first drilling will not be done until 1977 or 1978 outside Hammerfest (at 71° N). Nevertheless, from general geological indications it appears that possibilities for finding oil in parts of the northern shelf are quite good, perhaps as good or even better than they were in the North Sea. This conclusion coincides with statements from Soviet scientists claiming that the seabed in the Barents Sea and the Kara Sea (east of Novaya Zemlya) is rich in petroleum deposits. Until recently it was believed that the Russians were in no hurry to explore and exploit their Arctic continental shelf, but it now appears that they are speeding up their offshore programme. We know that their organization for Arctic geological exploration, the SEVMOR-GEO, is under orders to concentrate activities on the continental shelf and that the organization is in the process of expanding personnel and facilities. So before or from 1980 we may expect fairly rapid growth in the exploration of the Arctic continental shelf on both sides of the line dividing Norway and the Soviet Union. If oil is found in the early stages of this exploration, production may be expected to begin sometime in 1985–90.

If oil is found on the northern continental shelf in 1978 or soon afterwards, the Norwegian government will have to decide whether these resources are to be exploited immediately or if they should be left as a reserve to be used when the North Sea fields begin to dry up. If immediate exploitation is decided upon, this is bound to lead to a significant jump – perhaps a doubling – of Norwegian oil production by 1990. The reason for this is that with a North Sea development, including pipelines, plant and service facilities, as well as a regional economy that has been based on a production level of 90 million tons per year, it is hardly conceivable that production can be cut rapidly or even moderately reduced in the south as long as reserves are available in the area. At the same time, however, the high cost of operations in the north and the massive investments required in production, transportation and service facilities will make it mandatory to aim for a high level of production in the north as well. It is quite conceivable that for technical and economic reasons alone a production rate of 40–50 mill. tons per year is a minimum in the north and that a higher figure may be reasonable on other grounds.

In principle, production might be postponed in the north to coincide with a gradual reduction (or even phasing out) in the North Sea, but in practice this would be a most improbable development. For instance, it is hard to believe that the people in north Norway would accept such a delay when sufficient resources had been proven. The possibility of fields in the border zone, similar to the situation in the North Sea, could also force an early start and, of course, the existence of new, proven resources might lead to international pressures upon Norway to open them for more active exploitation. Finally, to postpone activities in the north until the depletion of the North Sea would mean that some time in the future – 50 years or more from now – a sudden shift would have to be made in the Norwegian economy, moving its base from the extreme south to the extreme north. This seems to be a highly implausible policy. The safest conclusion, therefore, must be that if oil is actually found in production quantities in the north, and if the market supports an acceptable price, the Norwegian output will increase substantially around 1985–90. It is not merely by chance that the government report on natural resources and economic development (Parliamentary Report No. 50 (1974–75)) bases one of its alternatives for economic development on the assumption that 'the production of petroleum will increase from about 90 million tons of oil equivalents in 1980 to 180 million tons in 2000'. What is important in our context is that the increase will come from the northern part of the continental shelf.

This is a possibility that attracts attention to the northern region and adds emphasis and urgency to the need to clear up any problems and issues that must be settled if oil activity is to proceed without undue disturbance. Such issues include, as was the case in the North Sea, an exact definition of the borderline of territorial waters, but it also involves questions about the application of the Svalbard Treaty and such questions as may arise in connection with the role of the Barents Sea as a passageway for the Soviet North Fleet.

SVALBARD AND THE BARENTS SEA

The Russians are known to prefer a solution to the Barents Sea problems through bilateral arrangements with Norway. In so far as the definition of the borderline is a question to be solved by the states involved, bilateral negotiations between Norway and the Soviet Union are the obvious procedure, and such negotiations are now being carried out. The Russians, however, have repeatedly indicated that they would like also to have talks at the highest political level to discuss the full range of questions relating to development in the Barents Sea area. There are several reasons why such talks and 'bilateralization' of relations in the Svalbard – Barents Sea area may not be tenable.

If the economic zone concept is internationally accepted and the Barents Sea divided into a Norwegian and a Soviet zone on the basis of an international convention, the two countries should not set up special rules, regulations and limitations between themselves on the use of the Barents Sea over and beyond those provided for in the convention. The purpose of the convention would be to define the rights and responsibilities of coastal states in their own zones and neither new rights nor new responsibilities should or could be established on a bilateral basis without consultation and consent from other states whose rights and privileges under the convention might be affected.

If, on the other hand, economic zones are established unilaterally, and not on the basis of an international convention, the problem becomes more complicated in that acceptance of the zone must be gained through negotiations with all parties concerned. The Norwegian government's earlier plans to extend national fishing limits to 50 miles have consequently meant negotiations with a dozen nations. In such negotiations any one party may insist upon special consideration and rights. However, if any one party does this, other parties too may feel that they have similar or equivalent interests and demand equal rights and considerations as a condition for accepting the zone arrangement. In other words such negotiations may be complex in any event and involve a number of parties.

In the special case of the Barents Sea the issue is further complicated by lingering doubts about the extent to which the Svalbard Treaty may apply to sea and seabed areas. In fact, any party which sees fit to claim that rights under the Treaty should apply to the sea and shelf around the islands may, in theory, claim to have an interest even in current borderline negotiations between Norway and the Soviet Union in so far as they concern drawing the dividing line between Svalbard and the Soviet islands. In practice such rather extreme claims can hardly be expected, though reservations made by the United Kingdom and the United States shortly before borderline discussions between Norway and the Soviet Union began in autumn 1974 may be seen as fair warning that these two parties are at least following the developments keenly and will assert their interests if they find this to be necessary. Indications are that several other countries will take a similar interest and regard the Svalbard Treaty as a legal basis for potential involvement in the northern development.

A peculiar result of the Svalbard Treaty of 1920 was that by recognizing 'subject to the stipulations of the present Treaty, the full and absolute sovereignty of Norway over the Archipelago of Spitsbergen' in Article I, the Treaty did in effect create a permanent basis for foreign interest in the region as well, and a potential for disagreement about the meaning and implications of the 'stipulations'. Though negotiated by the victorious powers of World War I and a select group of small countries (apart from Norway these were Denmark, the Netherlands and Sweden), the Treaty was and remains open for accessions by additional parties. At present some 40 states are contracting parties and in theory every state in the world may accede and so may express particular interest in the Svalbard region.

The most important of the Treaty stipulations was that:

'The nations of all the High Contracting Parties shall have equal liberty of access and entry for any reason or object whatever to the waters, fjords and ports of the territories specified in Article I; subject to observance of local laws and regulations, they may carry on there without impediment all maritime, industrial, mining and commercial operations on a footing of absolute equality.

They shall be admitted under the same conditions of equality to the exercise and practice of all maritime, industrial, mining or commercial enterprises both on land and in the territorial waters, and no monopoly shall be established on any account or for any enterprise whatever . . . (Article 3).'

Other provisions restricted the rights of Norway to levy 'taxes, dues and duties' (Article 8) and prohibited the establishment of naval bases and 'any fortification in the said territories, which may never be used for warlike purposes' (Article 9).

When the Treaty was negotiated there was no mention or awareness of offshore mineral resources, nor indeed was there any discussion of rights (national or international) beyond traditional territorial waters. With the current potential for development of seabed resources even under Arctic conditions and the introduction of the new principle of exclusive economic zones beyond traditional territorial waters, application of the Svalbard Treaty stipulations also in the sea surrounding the islands has become an issue. In the opinion of the Norwegian government the Treaty cannot be interpreted to include offshore areas such as a special Svalbard shelf or zone where Treaty conditions shall apply. Other governments have not shown much willingness to accept this opinion and the United Kingdom and the United States have explicitly reserved their position, thus keeping an option to dispute the Norwegian contention. Whether such a dispute occurs will probably not be decided on the basis of legal opinion alone but will involve careful consideration of strategic and security implications and, of course, of relevant economic factors. On the latter point, evaluation of general international supplies of energy – and Norwegian oil policy in particular – may be a decisive influence. If general supplies are difficult and declining, and other states feel strongly that Norway is contributing less than a fair share, they may be tempted to activate the Svalbard shelf issue.

Strategic as well as economic interests are involved in this issue. On the economic side, there is a question of access in the exploitation of the resources. In Svalbard, where the Treaty applies, nationals of any party to the Treaty have equal rights of access and participation, and the Treaty strictly limits the rights of the Norwegian government on taxation and fees. Compared with operations in the North Sea, operation under the Treaty implies substantial economic advantage. The possibility of equal access and the favourable tax rules are obvious reasons why some parties have reserved their position on the Norwegian attitude that the letter and principle of the Treaty are to apply only to the islands and their territorial waters and not to the sea and seabed beyond. In this connection we may also note that reservations about the extention of the Treaty may be used as a lever to press for an increase in Norway's oil production, or to secure access for national companies to participate in operations on the Norwegian continental shelf.

As far as economic interest and participation are concerned, therefore, there is a distinct possibility of a conflict of interest between Norway and her closest allies and partners. This could happen if a new energy crisis should occur and Norway persisted in maintaining a moderate level of production. However, the most likely situation in which Western powers might assert claims would be if any bilateral agreement were to be negotiated between Norway and the Soviet Union allowing for special rights and preferences for Soviet interests while excluding other parties. The Soviet Union, on the other hand, might see fit in some circumstances to accept and support the Norwegian interpretation of the limited applicability of the Treaty on certain conditions – such as non-participation by foreign firms or even special bilateral arrangements and preferences for Soviet interests. If this were to happen, the stage might be set for a conflict in which Norway would be supported by the Soviet Union against her own allies.

At the moment it is difficult to separate such yet theoretical arguments over legal points and economic interests from the strategic consideration that all major parties will and must take into account when they regard future development in the Barents Sea area.

The military importance of the Barents Sea is an unavoidable consequence of the fact that these are the waters where the Soviet Union has her safest year round access to the Norwegian Sea and the North Atlantic. Other naval routes from Soviet ports pass through such narrow straits as the Dardanelles and the Danish Belts, or, as with the Pacific ports, they are far removed from most vital centres. This is a main reason why the Northern Fleet, with its base in the Murmansk area, has become the largest of the Soviet fleets and why the bulk of the Soviet strategic submarine force operates out of the Kola bases. An additional advantage for the Soviet Union stems from the fact that the Svalbard islands are in effect a demilitarized zone and that a military threat from the islands is excluded. Because the Treaty allows all parties equal rights of access and economic activity in the islands, it is possible for any party to be continuously present and to observe on the spot how the islands are being used. The Russians are making full use of this right and today are the only nation apart from the Norwegians to maintain a permanent presence in the islands.

There are two Russian mining towns (Barentsburg and Pyramiden) on Spitsbergen Island and from these communities the Russians have every opportunity to observe activities in the whole region; for example, with the help of scientific parties travelling in the islands and through inspections carried out from helicopters. Such inspections, in the form of visits to scientific expeditions from other countries, occur regularly, often only two or three days after a party has established camp. At present, five Soviet helicopters are stationed in Svalbard and a permanent helicopter base is under construction at Cape Heer near Barentsburg.

The importance attached by Moscow to this preventive presence may be indicated by the fact that the Soviet population in Svalbard is twice that of the Norwegian population, while at the same time Norwegian coal production at Svalbard is larger than Russian production. At present, coal mining is the only economic activity, yet the Russians are now increasing their colony despite the fact that their coal reserves hardly warrant more intensive mining. The growth of the Soviet presence may be connected with Norway's plans to open a new mine (Sveagruben) and increase her population in the islands and, of course, with the fact that the opening of the Svalbard Airport in 1975, with scheduled commercial flights, facilitates a higher level of general activity in the Norwegian community in Svalbard.

We have every reason to believe that the sensitivity to change which the Russians are displaying in Svalbard is matched by a similar sensitivity to change in the Barents Sea and by a desire to forestall any development that may be regarded as detrimental to their strategic interests. Free and unhampered passage has long been a major concern of the Russians and was used by Molotov in 1944 as justification for his pressing the Norwegian foreign minister (Trygve Lie) for cession of Bear Island to the Soviet Union and joint Soviet-Norwegian government in Svalbard.

Today the Barents Sea is gaining added importance by being used as a deployment area for the new generation of strategic submarines. From positions in the Barents Sea, the new long-range missiles may reach any target in Canada and the United States. Unlike the earlier generation of submarines, therefore, the DELTA II does not have to leave the Barents Sea for targeting positions closer to the North American continent. Military experts will have to discuss the full strategic implications of this development, but we may assume that the Barents Sea is now more important in a strategic sense than ever before. We may also assume that with the deployment of an increasing share of their second-strike capability in the Barents Sea, the Russians will have an added reason for trying to establish and preserve their invulnerability in the area and that the United States, which is the primary target, will have an equal interest in neutralizing the Soviet force in the same region.

THE NORWEGIAN PREDICAMENT

All this may present problems for Norway. On the one hand, the country is about to become dependent on her northern continental shelf as a base for her continued economic development. On the other hand – and for the very reason that oil may soon be tapped from these northern resources – the area is coming more strongly into the focus of international attention, while at the same time the strategic balance between the superpowers may be more directly affected by the development in the same waters.

Norwegian policy, to the extent that it has been defined, is based on the assumptionthat full national control of the economic development on the Norwegian side of the dividing line in the Barents Sea is the best and indeed the only alternative. This will obviously be to the national economic advantage, but it is also assumed that it is the best way of avoiding rivalries for preference between other powers that may spill over into the strategic competition of the superpowers.

While proceeding on the assumption of the advantages of national control, Norway is at the same time becoming more vulnerable herself. Unless the guarantee presently provided through NATO is maintained, she will find it more difficult to defend the areas under her national control and will be more open to political pressure and blackmail. On the other hand, Norway's allies may insist that security has its price and press for a more determined contribution on her part to the western economy.

In this situation, and faced with the dangers of going it alone, while at the same time running the risk of actually exciting rivalry between the major powers if concessions are given to one side or the other, Norway may feel forced to actively seek arrangements that may unite rather than divide international interests in the region. In this respect an effort to establish a wider, Arctic scheme of cooperation may offer an attractive alternative. While dividing forces may still be strong on the Barens Sea issue, all Arctic powers have a community of interest in the general development of the Arctic region. Massive scientific research is still required in the north and this may be helped greatly by international co-operation. Svalbard, by offering equal rights of access and activity already and by having a central location in the Arctic, may provide an excellent base for joint scientific efforts. Operation in the Arctic, and particularly in the offshore areas, is a risky and difficult business with extreme demands upon personnel, equipment and techniques. In this respect too, co-operation, sharing of experience and joint development could be a great advantage. In one particular aspect the Arctic powers are mutually dependent upon each other: the Arctic is the most vulnerable environment on earth and the effects of extensive pollution will be far-reaching. For this reason any massive development in the Arctic will make an Arctic 'environment watch' mandatory. Finally, shipping in Arctic waters, when developed, will become an important trade. The establishment of navigational aids and provision for the security of ships and men, as well as the setting of joint standards and minimum classification requirements, is another area where co-operation may be helpful and indeed necessary and where Svalbard, because of its location and good communications, might offer special advantages – for example, for search and rescue work.

If the Norwegian predicament, which issues from the country's special geographical position and the concentration of diverse international interests in the Norwegian part of the Arctic, were to produce a policy that could result in more active international co-operation in the whole Arctic, this might indeed prove to be one of the more desirable aspects of interaction between the Arctic and the North Atlantic.

BJØRN BJARNASON

Iceland's Security Policy

It was a *conditio sine qua non* for the membership of Iceland in NATO that an Icelandic army would not be established, nor would foreign forces be stationed permanently in the country. Iceland is unique in the sense that she has never had her own armed forces. Only two years after the signing of the North Atlantic Treaty a special Defence Agreement between the United States and Iceland was made in the framework of the NATO Treaty. Since then those two matters – membership of NATO and the stationing of American forces in Iceland – have been subjects of great controversy, heightened at times by the seemingly unavoidable disputes with Britain and West Germany about fishing rights in Icelandic waters.

Since the implementation of the Defence Agreement in 1951 the termination of that agreement has twice been on the formal agenda of left-wing governments – 1956–1958 and 1971–1974. Internal discussion about Icelandic security is, of course, marked by the fact that the nation has no military tradition and no Icelandic specialists on military matters are to be found. The polemic is therefore based more or less on nationalistic and emotional sentiments.

In the first months of 1974 14 individuals collected signatures on a declaration to the Government and Althing (parliament) stressing support of NATO and opposing an untimely termination of the Defence Agreement. About 50 per cent of the electorate at that time – 55,522 voters – signed this declaration. A similar sentiment appeared in the municipal elections and the legislative elections in 1974. When a new government had been formed, a settlement was reached between the United States and Icelandic governments, leaving the 1951 Agreement intact but envisaging some changes at the Naval Base.

Any consideration of the attitude of the Icelanders to security matters must take into account the small size of the nation. It is a fact that small nations sometimes regard themselves as entitled to display more irresponsibility than large ones; because of their smallness they enjoy a special position. Little countries also tend to be more introverted than bigger ones and to think rather of their own immediate interests than of an equilibrium in international relations. In small countries there can be a complete *volteface* in foreign affairs when there is a change of government, which is almost unimaginable in larger states.

In 1918 Iceland became a sovereign state in monarchical union with Denmark. The legislation on which this union was based affirms Iceland's eternal neutrality in Article 19. This declaration was still in force when the country acquired full independence in 1944, and it has never actually been replaced by a declaration to the contrary, but denied only by Icelandic governmental action in security matters.

During World War II this declaration of neutrality was not strictly implemented. The country was occupied without conflict by Britain on May 10, 1940. On July 7, 1941 an agreement was reached between the Icelandic and the United States governments providing for the replacement of British by American troops. American troops then remained in the country until after the end of the war, and in October 1945 the United States government made a request to the Icelandic government for the opening of discussions between them concerning a long-term lease for military bases at three specified places in Iceland. The United States had in mind a lease for a period of 99 years, but in November of that year the Icelandic government replied that it could not agree to the talks. In September of the following year, 1946, the Icelandic parliament approved a resolution declaring that the agreement of 1941 had become null and void and that the United States should evacuate her armed forces from Iceland within six months. At the same time the United States was granted permission to use Keflavik Airport for a limited period so that she could fulfil the obligations she had assumed in connection with the military government in Germany. For this purpose the United States government was authorized to maintain such activities, equipment and personnel as might be necessary. This agreement was to remain valid as long as the United States was responsible for military government and control in Germany, but with the provision that after five years from the date of its signing either party was authorized to demand a re-appraisal. The agreement was approved in the Icelandic parliament on October 5, 1946, and it was still in force when Iceland joined NATO in 1949.

CONDITIONS FOR MEMBERSHIP OF NATO

The experience of the Icelanders during the war convinced them that neutrality alone would not prevent their involvement, especially if they had no means of defending that neutrality. With the advent of new military techniques, aircraft and submarines, the distance between Iceland and other countries no longer provided an adequate guarantee of security. Iceland's location half-way between the United States and the Soviet Union, the new military techniques, and tension in Europe combined in such a way as to invite a race between potential belligerents for facilities in Iceland, should an armed conflict break out in the Atlantic Ocean. When the Icelandic government was invited to become a founder member of NATO, it first ascertained what obligations Iceland would have to assume in the event of participation in the Alliance. Iceland's problem was that she was a country with no armed forces or defences of her own.

If a condition for participation were to be the establishment of a national military force or the stationing of foreign armed forces in Iceland in peacetime, the Icelanders considered participation inadvisable. A committee of three government ministers was sent to the United States to ascertain what the conditions for, and the consequences of, participation would be. In a report they drew up about their mission the following was stated:

At the conclusion of the talks it was declared on behalf of the United States:

1. that in the event of hostilities the Allies would request similar facilities in Iceland to those enjoyed during the last war, and that it would be entirely at the discretion of Iceland herself as to when such facilities should be granted;
2. that all other parties to the agreement fully understood Iceland's special position;
3. that it was recognized that Iceland had no armed forces and had no intention of raising an army;
4. that the stationing of foreign troops or bases in Iceland in peacetime was out of the question.

After these provisions had been approved, the Icelandic parliament passed a resolution proposing to authorize the government to become, on Iceland's behalf, a founder member of NATO. By joining the Alliance, Iceland affirmed her solidarity with the Western nations in a formal agreement and became more closely linked with her neighbours than before.

There is little doubt that Norwegian and Danish encouragement had a great effect on the attitude of the Icelandic Government, once the aforesaid conditions had been met. The policy of Norway and Denmark in security matters exerts a strong influence in Iceland, for Iceland's foreign policy has always been formulated with regard to the policy of the Nordic countries, particularly Norway and Denmark –

countries with which Iceland has for centuries had the closest contacts. During the years when NATO was established there was much talk of formal cooperation in defence matters among the Nordic countries. Although Iceland is one of the latter, there was never any question that she would participate in such cooperation. Iceland was not invited to take part in the discussions that were held between Sweden, Norway and Denmark. The problem of Iceland's lack of defence is not so urgent *vis-à-vis* the Nordic countries – or at least it was not so in those years – as it is with regard to the terms of the NATO agreement and the aforesaid conditions concerning Iceland's membership. It is completely within the power of the Icelandic Government to decide when recourse has to be had to special measures to ensure the security of the country. This was considered necessary in 1951, when the Defence Agreement with the United States was concluded. International affairs were then such that, especially because of the Korean war, there was a danger of an outbreak of general hostilities. A request was then received from NATO for a strengthening of Iceland's defences, as the facilities enjoyed in the country on the basis of the 1946 agreements were not considered adequate for its defence, that agreement providing merely for the right to use Keflavik Airport. The 1951 Defence Agreement between Iceland and the United States was made within the framework of the North Atlantic Treaty, as is stated in the Preamble:

'Having regard to the fact that the people of Iceland cannot themselves adequately secure their own defences, and whereas experience has shown that a country's lack of defences greatly endangers its security and that of its peaceful neighbours, the North Atlantic Treaty Organization has requested, because of the unsettled state of world affairs, that the United States and Iceland, in view of the collective efforts of the parties to the North Atlantic Treaty to preserve peace and security in the North Atlantic Treaty area, make arrangements for the use of facilities in Iceland in defence of Iceland and thus also the North Atlantic Treaty area. In conformity with this proposal the following agreement has been entered into.'

This Preamble expresses the twofold task of the US Defence Force in Iceland. On the one hand, it is to defend Iceland, and on the other to ensure the security of the sea areas around the country. Although there was more solidarity in the Icelandic parliament about the Defence Agreement than about joining NATO, disputes over the stationing of United States forces in Iceland since the signing of the Agreement have always been much more intense than those concerning membership of NATO. As may be seen from the final paragraph of the conditions imposed by Iceland for such membership, Iceland is not bound to accept armed forces in the country in peacetime; the interpretation of 'peacetime' has, however, been a matter of dispute.

Some indication of how this should be interpreted came to light in the disagreement that arose in 1956 about the stationing of the Defence Force in Iceland. A resolution was then introduced in parliament, proposing *inter alia*:

'Having regard to the changed circumstances since the 1951 Defence Agreement was concluded, and taking into account the declarations that there should be no armed forces here in peacetime, a review should forthwith be initiated of the organization then adopted, with a view to Iceland herself taking over the operation and maintenance of the installations, though not military duties, and to the evacuation of the armed forces. Should no agreement be reached on the amendment, the matter should be pursued by a notice of termination (of the Agreement) in accordance with Article 7 of the Agreement.'

This resolution was approved in parliament on March 28, 1956, when the members of the left-wing parties – that is, the Social Democrat party, Progressive party, Socialist (Communist) party and National Defence party – voted for it. The only party that opposed the resolution was the Independence party. There were two main points of view in the resolution. On the one hand, the armed forces were to leave because there was a state of peace; on the other hand, the Icelanders themselves were to take over the operation and maintenance of the defence installations, though without the formation of an army, for no Icelandic political party has had the creation of a domestic army on its programme. As a result of this resolution, talks began in autumn 1956 between the Government of Iceland and the United States on the implementation of the proposals – at the same time as the Soviet Army was suppressing the freedom movement in Hungary and war was erupting in the Middle East. The talks ended in an Agreement, which was ratified by an exchange of Notes on December 6, 1956, declaring *inter alia*:

'the governments of Iceland and the United States have therefore decided:

(1) that discussions pertaining to a review of the Defence Agreement, insofar as they concern an evacuation of the Defence Force, shall not be continued . . .'

In a report he submitted on the matter, the Minister for Foreign Affairs proposed that the talks with the United States should be suspended:

'In my opinion the world situation is now more serious than it was in 1951, when the Defence Force was brought to Iceland, and I consider that now is not the time to talk about sending the Defence Force away from Iceland, as matters stand. In my view, this would not only endanger the security of our country but, to no less an extent, invite danger – both for ourselves and for those other nations with whom we are co-operating in NATO.'

The government that conducted the discussions with the United States was composed of ministers from the Progressive, Social Democrat and People's Alliance (formerly Socialist) parties. The Independence party, which formed the Opposition, had from the beginning condemned the government's policy of terminating the Defence Agreement as irresponsible.

At this juncture, it is appropriate to explain briefly the attitude of the Icelandic political parties to the stationing in Iceland of the Defence Force and to security matters in general. The one party that has most consistently supported membership of NATO and not wished to take any steps that would impair the defence and security of Iceland, without careful consideration, is the Independence party (25 Members of Parliament, 42.7 per cent of the total vote in 1974). It was the Independence party that took the initiative about joining the Alliance and in the conclusion of the Defence Agreement with the United States. The party has never declared that a foreign defence force should be stationed in Iceland forever; but its spokesmen have expressed the opinion that times have changed since 1949 and 1951, and that there is no reason to adhere rigidly to the conditions then stipulated. In particular, there should be a review of the provision about 'peacetime' since the interpretation of that concept is bound to have changed, keeping pace with new military techniques. The Progressive party (17 Members of Parliament, 24.9 per cent of the vote in 1974), which is the second largest party in Iceland, supports membership of NATO but considers it desirable that the Defence Force should leave in stages, although NATO would be permitted facilities in Iceland for the purpose of checking on the movements of ships and aircraft in the Atlantic. The Social Democrat party (5 members of Parliament, 9 per cent of the vote in 1971) also supports membership of NATO. In 1956 it joined in the proposal for a review of the Defence Agreement, but now opposes measures that might unilaterally impair the country's defence and security.

The People's Alliance, a merger of the Socialist party and a left-wing splinter group from the Social Democrat party (11 Members of Parliament, 18.3 per cent of the vote in 1974) opposes membership of NATO and the stationing of the Defence Force in Iceland. However, the party has not made withdrawal from NATO a condition for its participation in government coalitions. The newest party, the Asso-

111

ciation of Liberals and Leftists and a splinter group from the People's Alliance (2 Members of Parliament, 4.6 per cent of the vote in 1974) seems in general to support membership of NATO, but there are certain factions within it that demand withdrawal from the Alliance and the expulsion of the Defence Force.

RECENT DEVELOPMENTS IN THE DEFENCE CO-OPERA-TION BETWEEN ICELAND AND THE UNITED STATES
During the years 1960–1970 the government in power consisted of a coalition between the Independence and the Social Democrat par-' ties. That government's policy in security matters was based on membership of NATO and on the Defence Agreement of 1951. It did not demand a termination of the Agreement, though it kept Iceland's defence needs continually under review (and called in some foreign experts for that purpose). In July 1971 there was a change of government in Iceland and that which came to power contained ministers from the Progressive party, the People's Alliance and the Association of Liberals and Leftists.

Although security matters had not been a matter for dispute in the election campaign, the new government formulated a new security policy and its programme stated, for example:

'The Government considers that efforts should be made for a relaxation of tensions in the world and the strengthening of peace and reconciliation by closer contacts between nations and by general disarmament. It considers that peaceful relations between nations will best be preserved without military alliances. There is disagreement between the Government parties on the attitude towards the membership of Iceland in the North Atlantic Treaty Organization. However, under unchanged conditions the present order shall continue, but the Government will endeavour to follow as closely as possible developments in this field and re-evaluate at any time the position of Iceland in accordance with changed conditions. The Government agrees to the convening of a special conference on European security. The Defence Agreement with the United States of America shall be taken up for review of termination for the purpose of having the Defence Force leave Iceland step by step. The aim shall be to have the departure of the Defence Force take place during the electoral term.'

According to Article 7 of the Defence Agreement, either government may – after previous by notifying the other – at any time request the Council of NATO to ascertain whether the facilities in question are any longer necessary and to make proposals to both governments on whether the Agreement should remain in force. If such a request for review does not result in agreement being reached between the two governments within six months of the submission of the request, either government may subsequently terminate the Agreement, which shall become invalid twelve months later. It should be mentioned that when the Defence Agreement was to be terminated in 1956, the matter was referred to the North Atlantic Council on the basis of the relevant provision in the Defence Agreement. The Council's recommendation was clearly stated in the following paragraph:

'The North Atlantic Council, having carefully reviewed the political and military situation, finds a continuing need for the stationing of forces in Iceland and for the maintenance of the facilities in a state of readiness. The Council earnestly recommends that the Defence Agreement between Iceland and the United States of America be continued in such form and with such practical arrangements as will maintain the strength of the common defence.'

On June 25, 1973 the Icelandic government referred the review of the Defence Agreement to the North Atlantic Council for the second time since 1951. The Council finished its study in December 1973, stating that the review of the Defence Agreement, while giving all possible consideration to the wishes of the Icelandic government, should ensure the continuation of this Agreement in such a manner that the use of the Icelandic facilities would contribute to the maintenance of Alliance security as a whole, and particularly to securing a crucial link between the two sides of the Atlantic.

After January 1, 1974 it was in the hands of the Althing to terminate the Agreement. In January the Icelandic Foreign Minister had talks with the United States Foreign Minister and officials in Washington – an informal step in the general study being undertaken.

Preparations for formal talks were continued while the matter was under study in the NATO Council. On March 13, 1974 the Icelandic Government formed a mandate for the Foreign Minister, who tabled the Icelandic proposals in Washington on April 8 and 9, 1974, when he met United States officials formally, to discuss the revision of the Agreement. Those proposals or the draft basis for discussion were the following:

'1) The Defence Force which is now stationed in Iceland shall be withdrawn gradually from the country. The withdrawal shall be so effected that before the end of 1974 one-fourth of the force will have been withdrawn, one half before the middle of 1975, three-fourths before the end of 1975 and the remainder before the middle of 1976.

2) In order to comply with the obligations of Iceland towards NATO the Icelandic Government proposes the following procedure:

a. Aircraft operating on behalf of NATO shall have landing rights at the Keflavik Airport whenever necessary for surveillance flights in the North Atlantic. There will not, however, be a permanent base for aircraft here. This matter to be further dealt with through negotiations.

b. For the purposes of such landings NATO will be authorized to station at Keflavik airport a group of civilian technicians for the maintenance of the aircraft mentioned above. The number of this personnel to be negotiated, but with the view that the operations will be on a limited scale.

c. After the withdrawal of the Defence Force, Icelandic nationals shall take over law enforcement at the airport. Icelandic nationals shall before that time receive training in the specialized law enforcement duties required.

d. Iceland shall make available the required manpower for the necessary staying at Keflavik Airport in accordance with the above provisions.

e. Iceland will take over the operations of the radar stations on Reykjanes and Hornafjordur, when trained Icelandic personnel is available.

f. Civil aviation will be completely separated from the operations which in accordance with the above will take place at Keflavik Airport in compliance with the obligations of Iceland to NATO.'

These proposals were never put to the vote in the Althing; they were introduced to the Foreign Relations Committee. In his annual report on foreign policy delivered in the Althing in April 1974 the Foreign Minister said that at the meeting with United States officials in Washington on April 8 and 9 it was decided that the Americans would consider his proposals and deliver a formal answer later. Then the Foreign Minister said:

'It is my view that we should try to solve this matter through negotiations as long as we can, before the step will be taken to terminate the Defence Agreement of 1951, and this matter is under study in Washington and here.'

The left wing coalition government was facing dangerous economic problems. Internal disagreement in the coalition led to the dissolution of the Althing and new elections on June 30, 1974. The Independence party was the indisputable victor, increasing its percentage by 6.5 per cent – from 36.2 to 42.7 per cent. The party's victory was not least due to its firm foreign policy line, stressing the need for NATO membership and continued defence cooperation with the United States.

On August 28, 1974, after two months of negotiations, a new government was formed – a grand coalition between the Independence and Progressive parties, supported by 42 out of 60 members of the Althing and 67.6 per cent of the voters. The government is lead by the Independence party but the Progressive party maintains the post of foreign minister. The policy statement of the new government contained the following about security matters:

'To ensure her security Iceland will continue her membership of the North Atlantic Treaty Organization. Special co-operation shall be maintained with the United States of America as long as a defence and reconnaissance base is operated in this country under the auspices of the North Atlantic Treaty Organization.

Negotiations on the organization of the defence of the country shall be continued with the aim of enabling the Keflavik base to carry out its functions in accordance with the security interests of Iceland at any time.

It is the policy of the Government that non-military functions of the Defence Force be taken over by Icelanders. All measures towards this end shall be accelerated as much as possible.

Members of the Defence Force shall be housed within the limits of the base area, as soon as conditions permit. The Defence Force activities and civil airport operations at Keflavik Airport shall be separated.'

On September 26, 1974 the Foreign Minister again had talks with United States officials about the Defence Agreement. He then announced that the government had formed a new policy and would not set time limits for the departure of the Defence Force. The government had reached the conclusion that, for the time being, necessary changes would be made within the framework of the Agreement of 1951. The result of those talks was an Exchange of Notes on October 22, 1974. Here it is stated that the two governments: 'Agree that the present situation in world affairs as well as the security of Iceland and of the North Atlantic Community, call for the continuation of the facilities and their utilization by the Icelandic Defence Force under the Agreement on mutually acceptable terms.'

In the Memorandum of Understanding that was issued with the Exchange of Notes it is stated that the United States will seek to reduce her military personnel by 420, to be replaced by qualified Icelandic personnel. The United States will seek to construct family housing units on the base area to accommodate eligible US military personnel stationed in Iceland. The civilian air terminal will be separated from the base facilities. The formal review of the Defence Agreement was concluded by this Exchange of Notes. The Agreement can now be terminated by either government by 18 months notice as before.

The implementation of the Memorandum of Understanding has already started. There have been some reductions in US military personnel and new housing is under construction. The most sensitive problem, which will be discussed in the wake of the enforcement of the Memorandum, is the competition for labour between activities on the base and the fishing industry of neighbouring communities.

PREMISES ON WHICH THE POLITICAL ATTITUDE IS BASED

It might be expected that the attitude of the Icelanders to the defence of their country would be motivated primarily by security interests, and this is indeed the case among those taking account of military developments around the country, as well as Iceland's situation in the Atlantic Ocean. But those who demand a termination of the Defence Agreement with the United States for nationalistic reasons are much more conspicuous and vociferous.

In other countries it is either concern about the high cost of maintaining suitable defences, or else a desire for peace that is the main reason for going slow on strengthening their armed forces; in Iceland it is much more a question of national pride. Opponents of the Defence Agreement with the United States assert that it curtails the nation's independence and that the stationing of the Defence Force in Iceland will have a dangerous effect on Icelandic culture. On the former point, it can hardly be said that an agreement on defence made on a basis of equality with another nation constitutes an abridgement of national independence, yet, on the other hand, it is a defect in the independence of a nation if she is unable to defend herself. As far as undesirable cultural influence is concerned, all Icelandic governments since 1951 have done their utmost to prevent the presence of American troops having any such effect on the national life. Strict regulations have been introduced controlling movement outside the base – so strict in fact that it is unlikely any nation other than America would tolerate them. At one time television from the US Base could be seen in the most thickly populated area of Iceland, but after some sharp disputes this was limited, as far as technically possible, to the confines of the base itself. The American involvement in Vietnam was, of course, harmful for their image in Iceland, as elsewhere, and this fact was duly exploited by the opponents of the base in Keflavik. One should not forget either that the policy of detente does not necessarily support the importance of continuing defence co-operation. As this policy undermines the will of many Western countries to contribute sufficiently to their defences, it also undermines the will of many Icelanders to defend the stationing of foreign troops in their country.

Although particular cases of irritation have often been the cause of much debate and argument about whether there should be any foreign

116

defence force in Iceland, in the final analysis it must surely be security interests that decide whether the US forces should be asked to leave or not. In fact, that has always been the policy of the Parliament and the Government in Iceland so far.

ECONOMIC ASPECTS

Some states lease their land to foreign powers for military use. The question has been asked whether the same might not apply to Iceland, i. e. that the US forces should be allowed to remain if Iceland received adequate financial remuneration in return. There has always been much discussion about the influence of the Defence Force and its activities on the Icelandic economy and labour market. Although the financial aspects are of course important, they have not been decisive. It has been Icelandic governmental policy that the independence and security of Iceland and her contribution to the defence of world peace should be the main criteria as to whether the Defence Force should be stationed in the country or not. Shortly after the force's arrival there was some unemployment, and thousands of Icelanders then went to work for the Americans and on the latter's building projects. If the Icelanders themselves had not desired this, there would have been no need to employ a single Icelander on such work. Today, unemployment is merely a local and seasonal phenomenon in Iceland.

About 800 Icelanders are now working for the Defence Force, and many of them could undoubtedly continue to be employed in connection with the operation of Keflavik Airport even if the Defence Force were to be withdrawn. Net receipts from the Defence Force according to balance of payments statistics published by the Central Bank of Iceland amounted to 16.5 million dollars in 1972, 20.5 million dollars in 1973 and 22.1 million dollars in 1974. In 1972 the revenue received in connection with the Defence Force amounted to 2.2 per cent of the Icelandic Gross National Product, 2.0 per cent of the 1973 GNP and 1.7 per cent in 1974. If it has really been Iceland's intention to profit by the stationing of the Defence Force there, she would almost certainly have begun by imposing tariffs on its imports, but in fact the US military authorities are allowed to import duty free military equipment, together with a reasonable amount of provisions, stores and other goods destined for the troops, their families, American contractors and their employees who are not Icelandic citizens.

THE SECURITY ASPECTS

Only the Icelanders are competent to make decisions on all these matters. They themselves must decide whether they want the Defence

Force in their country or not, from the point of view of their own interests. They can reject it for cultural or nationalistic reasons, and they can form an opinion on whether their economy could withstand its withdrawal or not. It is more difficult for them to decide what the military or strategic consequences might be if the Defence Force were to leave Iceland. As mentioned before, Iceland has no military experts in her service, though it is a basic duty of the Icelandic government, as it is for the governments of other countries, to ensure national security at all times and in the best possible way. The lack of such experts is certainly a serious matter. One of its effects is the possibility of expert reports on the military or strategic importance of Iceland appearing suspect, insofar as these come from non-Icelanders and may, therefore, be said to serve the interests of other nations.

On the other hand, the security interests of the countries bordering the North Atlantic are so interconnected that it is hardly possible to draw a distinction between them. The Icelanders realize this, and it is fairly certain that the majority do not wish to impair the security interests of their neighbours by unilateral action, or to imperil the state of peace that has prevailed in the Atlantic since the end of World War II.

As stated above, the concept 'peacetime' is the key word in the arguments of those who otherwise desire co-operation with the Western countries when they consider whether there should be continued co-operation in defence matters with the United Sta¹ Is the reason why that concept has become a reality the fact that there is not a state of war in Europe? When the Defence Agreement was concluded in 1951, particular reference was made to the Korean war on the other side of the world. When the termination of the Defence Agreement in 1956 was rescinded, particular reference was made to the invasion of Hungary and the war in the Middle East. And in the Exchange of Notes on October 22, 1974 is stated 'that the present situation in world affairs, as well as the security of Iceland and of the North Atlantic community' calls for the continuation of the Defence Agreement. It must also be borne in mind that, when this 'peacetime' concept was included in the conditions for Icelandic membership of NATO, the Atlantic States had a near monopoly on sea power. Early in the last decade, however, the situation was completely changed when the Soviet Union commenced her great expansion in the Atlantic,so that now it may be said that the areas east and south-east of Iceland have become a regular training ground for Soviet warships.

The 'peacetime' provision was introduced when it was possible to anticipate the outbreak of war in advance. Now, however, hostilities may break out without warning and could result in Iceland being confronted with a *fait accompli*: she could be occupied before NATO was able to send military forces there, assuming that defence forces were not already on the spot. It is likely that NATO would not send such forces, unless the Organization believed the situation to be very critical, for such action could lead to a general war. NATO might come to the conclusion that non-intervention would be preferable to the risk of general war. The danger of such a *fait accompli* would be much more acute if United States armed forces were to leave Iceland.

There are two tasks for the Defence Force in Iceland: to defend the country itself, and to ensure the security of the North Atlantic. The latter aspect is so important for, for instance, Norway, Denmark, Britain, the United States and Canada, that these countries would almost certainly make considerable sacrifices to secure a friendly attitude on the part of Iceland and an assurance that she would not constitute a threat to them. These are the five countries with which Iceland has generally the most contact: with Norway and Denmark because of traditional ties and kinship; with Britain and the United States because of friendly relations and commerce; and with Canada because large numbers of people of Icelandic descent live there.

JOHAN JØRGEN HOLST

Prospects for Conflict, Management and Arms Control in the North Atlantic

It seems useful at the outset to distinguish between the North Atlantic as an arena of conflict and as a source of conflict between states. Obviously, any such distinction will be blurred in particular instances, but as a general principle of classification it seems useful. Furthermore, it is not possible to discuss the prospects for conflict in terms of pro- bability estimates. The analysis must focus instead on identification of distinctive kinds of conflict, and the *conditions* which may determine their eruption and course.

THE NORTH ATLANTIC AS AN ARENA OF CONFLICT

The North Atlantic provides the vital connecting tissue for the North Atlantic Treaty Organization. The credibility of the system of guarantees upon which the alliance is based would crumble in the event of loss of control over the ocean areas which separate and con- nect North America and Western Europe. The North Atlantic remains the vital transmission belt for the projection of American power into the European scene, and a major war in Europe would convert the North Atlantic into a vital arena of conflict. The extent to which the battle for sea control and sea denial would be protracted would depend in large measure on the pace of a war in Europe and on the territorial advances of the opposing armies.

The immediate challenge is unlikely to assume the familiar and un- ambiguous shape of a clear-cut attempt at hegemonial conquest. The potential sources of conflict in Europe today are not limited to those which derive directly from the East–West division. The major sources of instability are rooted in the socio-political instabilities of the polities of southern Europe and the prospective occurrence of a series of suc- cession crises. The process of social transformation can produce acute tension between the domestic political order and the structure of secu- rity arrangements in Mediterranean Europe. Such tensions could in some circumstances produce pressures and incentives for unilateral or

120

competitive external intervention in the political processes of the countries of southern Europe. Chains of events could produce a rapid erosion of the atmosphere of détente and even involve the two dominant alliances in direct confrontation. The differential impacts on the security positions of the various members of the North Atlantic Alliance could make it very difficult to obtain consensus on appropriate measures of response. Naval power demonstrations could be designed to maximize differentiation of the risk assessments. Similar considerations apply to intra-alliance conflicts which might emerge in the wake of another war in the Middle East. Such processes could result in a serious weakening of the strategic position of the Atlantic Alliance with respect to its ability to establish sea control in critical regions of the North Atlantic. The Azores are probably the most important pieces of real estate in this connection. In the event of war, trans-Atlantic convoys would probably select transit routes south of the Azores in order to maximize distance to Soviet submarine and air bases on the Kola peninsula. Such a choice would, however, depend on NATO's continued ability to seal the exits from the Mediterranean. Bases in France and Portugal would remain very important for providing defensive air coverage in the Eastern portions of the Atlantic.

Under conditions of disarray in the Atlantic Alliance, naval power could be invoked deliberately for its power of suggestion. Demonstration of the power to interdict the sea lines of communication and deny the establishment of sea control in key ocean areas could seriously erode the perceived credibility of the trans-Atlantic guarantees of assistance. The shadow of force could weigh rather heavily on an environment of frustrated hopes for détente and conflicting approaches to the volatile relations between state and society in southern Europe. The real issue here is not whether a long or a short war in Europe is more likely; a combination of political transformation in southern Europe and the demonstration of improved naval prowess on the part of the Soviet Union could curtail Western options with respect to protracted warfare in Europe. Such curtailment could in turn affect assessments and behaviour during major crises.

The Soviet Union possesses a rather potent power of naval interposition, particularly in the north-east Atlantic. The large submarine fleet and the naval air arm constitute a serious threat and obstacle to the projection of American power ashore in northern Europe during an emergency. The Soviet surface navy is probably no match for the American Navy in a prolonged battle for sea control, and it is unlikely to be deployed for such purposes.[1] However, its limited staying power, combined with its high-speed capability for reaching the GIUK gap before the US Second Fleet, provide Moscow with bargaining leverage and escalation deterrence in a north European crisis. The power of interposition is likely to influence the calculation of risk during a crisis and could cause serious delays in any crisis intervention – for example, in north Norway. However, the Soviet ability to threaten the transfer of an American Marine Amphibious Force to north Norway might motivate a rapid airlift of American troops to Norway for the purpose of confronting the Russians with a serious escalation threshold on the ground in Norway. Transfer to south Norway may seem preferable in the pre-war phase of an emergency, for the purpose of avoiding direct confrontation and inserting another rung in the escalation ladder. The rapid deployment of long-range shore-based air power to airfields in south Norway and Keflavik, Iceland, would probably constitute the most potent short-lead-time sea-denial option available to the United States in a crisis situation in the northeast Atlantic.[2] Thus, crises erupting in northern Europe may be dominated by a competition in risk-taking and the rapid implementation of commital moves, in order to transfer the onus of escalation to the shoulders of the adversary. It is possible to envisage also a kind of prolonged *drôle de guerre* in which the moves resemble shadow-boxing manoeuvres. The superior staying power of American surface units are likely to prove advantageous in such circumstances.

The transfer of American, British and Canadian forces to Norway would not necessarily take place in the context of a direct threat of attack against Norway. The needs would vary with circumstances. Thus, we should distinguish between reinforcements during an emerging crisis, after an attack on Norway, in response to an attack or pressure on another country in the region (such as Finland), and in the context of a general war in Europe. The need for substantial fighting troops with heavy equipment would be greatest in the second and fourth cases. Such emergencies would tend to remove many of the political obstacles associated with American naval forces shooting their way into the Norwegian Sea for the purpose of projecting power ashore in Norway. In principle, heavy equipment could be pre-stocked in Norway as a means of circumventing the interdiction powers of the Soviet Navy. This may be a more acceptable solution for Canada, which does not have global responsibilities but does have certain forces earmarked for intervention in the AFNORTH area. For the United States pre-stocking could mean a loss of flexibility with respect to future intervention options, though the REFORGER concept, based on double sets of equipment, would not carry the same potentially adverse implications. Moreover, it could produce serious political tensions between Norway and the United States if the latter wanted to draw on the stocks to support the operations in a conflict outside Europe about which political assessments within NATO varied sharply, as they did during the wars in south-east Asia. British Royal Marine Commando Groups, however, would probably require pre-stocking only of over-snow vehicles and some helicopters.

The risk calculus and behaviour in a crisis in the north-east Atlantic will be influenced, *inter alia*, by the pre-emptive instabilities inherent in the Soviet need for surge deployments through the potential choke points. The pace of events may be dominated also by a Soviet need to deny bases in Norway and Iceland to the North Atlantic Alliance. But such dramatic moves would signal intentions of a major reach for hegemony in Europe, and they would presumably cause dramatic counter-measures in central and possibly also southern Europe. It seems more likely that Soviet naval power will serve the political function of complicating the Norwegian security calculus in the tension field between deterrence and reassurance. Uncertainty and a heightened sense of risk during crises could become dominant elements in the expectations and political assessments in northern Europe, thereby weakening the coherence and cohesion of the Western security structure.

It is possible to envisage a kind of 'grey war' at sea, waged primarily by submarines, with the sinking of adversary submarines being undertaken to demonstrate prowess and daring as well as to weaken the morale of the adversary. Such actions could also be undertaken so as to prevent the collection and transmission of certain kinds of intelligence information. Similar and less bizarre operations by submarines in the territorial waters of the littoral states could be used either to demonstrate such states' incapacity to deny access, or else to collect data about the submarine environment in potential hiding areas or missile-firing positions. Under such circumstances coastal-state search operations could disrupt the missions and cause abortions, even if they were not able to force the submarines to surface (which is not invariably a necessary, or indeed desirable objective). However, such hide-and-seek operations in peacetime could produce situations in which the parties involved became locked into a major confrontation, either by default or inadvertence.[3]

Some 70 per cent of the Soviet SSBN force is home-ported with the Northern Fleet on the Kola peninsula. The bulk of the Soviet SSBN force consists of Y-class boats (some 25 with the Northern Fleet) carrying the SS–N–6, which has a range of 1,500 nm. The missile's limited range imposes the need to operate in the western half of the Atlantic, and therefore to transit through the GIUK gap. For several reasons which are not very well understood, only a rather small portion of the Y-class SSBN are kept on station at any time. (Apparently only

123

two of the Northern Fleet's 25 Y-class boats are on station at any time south of the GIUK gap). An intense crisis could therefore generate strong pressures for a surge deployment through the gap. This state of affairs would in turn act as a strong stimulus to the rapid establishment of anti-submarine barriers across the exits. Such barriers would presumably consist of a combination of fixed- (SOSUS) and towed- (SURTASS) array sensor systems, maritime patrol aircraft, helicopters, SSN and surface vessels; the deployment of deep barriers of 'smart' mines of the encapsulated torpedo (CAPTOR) variety may constitute a particularly potent option in future. The race for exit and closure could cause rapid escalation.

The D-class SSBN's carry the 4,200 nm SS–N–8, thus eliminating the need to pass through the GIUK gap and permitting rearward patrols in the Greenland Sea, the northern portion of the Norwegian Sea and the Barents Sea. In order to mount first strike threats against time-urgent targets like airfields, radars and command centres, the SSBN would presumably move closer to the targets, but, in a situation in which the Soviet Union has the option of rearward deployments, substantial passage through the GIUK gap could signal a first-strike threat. Surveillance of the north-east Atlantic would consequently remain a significant American interest.

Apparently some two of the 15 D-class submarines with the Northern Fleet are normally deployed to patrols in the Barents and Greenland Seas. It is too early, however, to predict the future area and volume of patrols; much will depend on whether Moscow opts for maintaining a higher portion of the total force on patrol than at present. The Barents Sea is rather shallow and may not provide D-class boats with enough ocean space at optimum depth to sustain a substantial deployment. Previous missile-carrying submarine classes were tried out in the Barents Sea prior to their commitment to forward deployment, and exclusive deployment there would constitute a rather extreme concentration of high value targets in a limited space, encouraging a focused multidimensional ASW effort on the part of the western powers. Moreover, it also does not seem likely that Moscow would choose to commit its SLBM's to the same attack corridors as the land-based missile systems if it wanted to maximize deterrence. Thus, it would seem likely that the D-class SSBN deployment area will comprise a wide arc from the Barents to the Greenland Seas. Some patrols may extend also to the South Atlantic. Such a mode of operations would increase flexibility of target coverage and avoid first-strike connotations being associated with exits through the GIUK gap.

It now seems likely that Moscow will maintain a force structure which combines long and medium range submarine launched missile forces also in future. Such a composition will enable the Soviet force to threaten North America from different directions and across different ranges, thus complicating the problems of active defense. The mixed posture is evidenced by the flight testing of successor missiles to both the SS–N–6 and SS–N–7 missiles, the solid propellant SS–NX–17 and liquid propellant SS–NX–18 respectively.

THE NORTH ATLANTIC AS A SOURCE OF CONFLICT

In the preceding section reference was made to several structural instabilities of the naval dispositions in the north-east Atlantic. The Soviet fleet suffers from vulnerable base concentrations and the need for rapid dispersal and exit through choke points in a crisis. An intense crisis could be accentuated by a race for key strategic nodes like the GIUK gap, and by incentives for pre-emptive attacks against air bases in Norway and Iceland. The environmental instabilities could be compounded by the one-shot posture of the Soviet surface navy, which has a very limited reload capability. As well as the actors, the structural characteristics of the arena would have a significant impact on the course of a conflict, but they are unlikely to constitute independent sources of conflict.

The territorial *status quo* in the North Atlantic area has not been challenged in the post-war period. The constellation of conflict has been derived from the East-West confrontation in Central Europe and the evolution of the central balance between the two superpowers. However, competition for access to ocean resources could in future generate novel sources of conflict.

Even if the United Nations Third Conference on the Law of the Sea (UNCLOS III) approves a treaty establishing the principle of 200-mile exclusive economic zones (EEZ), the process of implementation and regional adjustment could produce conflict. However, it seems more likely that UNCLOS III will not be able to muster the required plurality for adoption of a comprehensive treaty; in that event we shall see a series of unilateral actions on the part of the coastal states. These actions may take place in a regionally concerted fashion. This looks the most likely scenario in the North Atlantic, involving primarily the United States, Canada, Denmark and Norway. (Iceland, with its extreme dependence on fisheries, is a case *sui generis*). Concerted action is, however, dependent on the ability of national governments to resist local pressures for rapid unilateral action; should the dyke break in one place the most likely result would be a scramble and the failure to reach a negotiated outcome.

125

The establishment of EEZ's will give the coastal states extensive regulatory powers over conservation of fish stocks. However, according to the Single Negotiating Text which constitutes the basis for the next set of negotiations in UNCLOS III, the coastal states are obliged to permit optimum exploitation and to admit outsiders to harvest the difference between the maximum sustainable yield and the coastal state's harvesting capacity. Disputes may arise here between the coastal state in question and the other members of the regional fisheries commissions – ICNAF (International Commission for the Northwest Atlantic Fisheries) and NEAFC (North East Atlantic Fisheries Commission) – over the determination of the maximum sustainable yield. Estimates are bound to differ, and experience indicates that estimates, as well as the total sum and allocation of quotas, are reached through a political bargaining process. Such conflicts could become accentuated as primary and residual regulatory powers become vested in the coastal states. The interested parties are likely to include demands for access to fish in their general negotiating posture in trade and security fields, thus enabling them to mobilize bargaining leverage across issue areas. It is clear that the establishment of EEZ's affects allocation as much as conservation. Disputes will arise over who gets what, when, and how. Such disputes will be waged also with specific claims to historical rights, traditional access, and equitable allocation.

The pattern of potential conflict will not necessarily conform to the one that has structured the security system in the North Atlantic area. However, we should note that the major non-littoral trawling states in the North Atlantic include the Soviet Union, GDR, Poland, Rumania and Bulgaria. The fish catch from the north-east Atlantic is more than twice that of the north-west Atlantic. Tables 1–3 below provides some clues for predicting the possible pattern of conflict over access to various fishing grounds. It should be noted, of course, that there are important variations from year to year. Furthermore, regulations and closures in some areas will transfer the activities elsewhere; thus there has been an important shift in recent years from the north-west Atlantic to the Norwegian and Barents Seas. Iceland's extension of her fishing zone to 200 miles has caused similar transfers. Herein lie some of the seeds of potential conflict arising out of unconcerted unilateral actions.

We note that four countries – Norway, the Soviet Union, Britain and Denmark – accounted for 62 per cent of the catch in the north-east Atlantic in 1973, while three countries – the Soviet Union, the United States and Canada – took 75 per cent of the fish in the north-west Atlantic. Allocation in the Barents Sea and around Svalbard is primarily a matter involving Norway and the Soviet Union, with Britain and Poland as the major 'outsiders'. Around Iceland the big

Table 1: Nominal Fish Catch (all species) in the North-West Atlantic, 1973.

	Million metric tons	% of total nominal tons
Canada	930.3	20.7
Denmark	0.5	—
Faroes	26.4	0.6
France	36.4	0.8
E. Germany	185.2	4.1
W. Germany	94.4	2.1
Greenland	44.0	1.0
Iceland	0.0	—
Norway	70.9	1.6
Poland	255.1	5.7
Portugal	124.3	2.8
Spain	180.6	4.0
USSR	1.357.4	30.3
Britain	8.4	0.2
United States	1.066.8	23.8
Total Nominal Catch	4.485.0	100

Source: Calculated from *FAO Yearbook of Fishery Statistics,* Vol. 36, 1973, *Catches and Landings* (Rome: Food and Agricultural Organization of the United Nations, 1974), p. 565.

Table 2: Nominal Fish Catch (all species) in the North-East Atlantic, 1973.

	Million metric tons	% of total nominal catch
Belgium	52.7	0.5
Denmark	1.450.0	12.9
Faroes	220.0	2.0
France	659.1	5.9
E. Germany	161.0	1.4
W. Germany	364.0	3.2
Greenland	0.3	—
Iceland	906.2	8.0
Netherlands	340.8	3.0
Norway	2.786.6	24.8
Poland	215.7	1.9
Portugal	277.4	2.5
Spain	674.1	6.0
Sweden	216.3	1.9
USSR	1.611.1	14.3
Britain	1.136.0	10.1
Total Nominal Catch	11.235.0	100

Source: *Ibid*. p. 566.

Table 3: Percentage of Total Nominal NEAFC catch from important North-East Atlantic fishing areas, 1972.

	North Sea	Norwegian Sea	Barents Sea	Spitzbergen/ Bear Island	Faroes Plateau/ Bank	Iceland Grounds	East Greenland
Denmark	37.3	—	—	—	—	—	—
Faroes	3.9	—	—	0.5	27.3	1.7	1.9
France	3.4	0.1	2.5	—	30.3	—	—
W. Germany	2.7	1.6	0.3	0.4	10.3	9.7	60.5
Greenland	—	—	—	—	—	—	0.6
Iceland	1.1	—	—	—	—	67.9	18.0
Norway	18.8	84.8	49.8	14.3	7.5	0.3	—
Poland	0.1	0.3	0.2	5.7	—	—	17.8
Spain	—	1.0	—	—	0.9	0.1	—
Britain (Eng-land/Wales)	5.5	1.8	4.8	5.9	6.7	18.6	0.5
Britain (Scotland)	9.0	—	—	1.7	17.1	0.4	—
USSR	5.4	0.2	42.4	71.4	—	0.1	0.7
Area catch as % of total NEAFC catch	29.6	18.2	12.3	1.0	1.2	9.9	0.5

Source: Calculated from Bulletin Statistique des Pêches Maritimes, Vol. 57,
(Copenhagen: Conseil International pour l'Exploration de la Mer, 1974), pp. 16–17.

'outsiders' are Britain and West Germany. In the north-west Atlantic the dominant 'outsider', the Soviet Union, outweighs even the United States and Canada; other important 'outsiders' are Poland, East Germany, Portugal and Spain. The EEC countries took 86.9 per cent of the Community's 1973 catch in the NEAFC area, and only 4.2 per cent in the ICNAF area. Great Britain took a third of its catch from the waters of non-member states, mainly Norway and Iceland.

The establishment of 200-mile EEZ's will produce a disparity between legal rights and policing structures: policing the enormous ocean areas in question will be beyond the capacity of the smaller coastal states, like Denmark, Iceland and Norway. Random patrols could be employed to maximize detection, but enforcement would basically depend on the co-operation from flag-states.[4] A major battle for access to fishing resources could involve naval confrontation – and not only between NATO countries and the Soviet Union, but also between NATO allies. Intra-NATO conflicts would tend to reduce the credibility of NATO support in subsequent disputes with the Soviet Union. Cutting trawls, ramming, etc. are among the measures available to fishery inspection vessels, but such tactics are effective only in the absence of armed escorts. Compact and cheap precision-guided missiles, as well as cannon-launched guided projectiles, may to some extent alter the power balance between large navies and coastal states attempting to deny access in limited and particular confrontations. In principle we could envisage confrontations where the trawlers carry missiles.

Disputes and conflict may arise also with respect to regulatory powers over river-spawning fish, such as salmon. Denmark is the odd-man out on this issue. The Single Negotiating Text in UNCLOS III gives regulatory powers to the states in whose fresh water these fish originate.

The incentives to reach negotiated agreement on access and enforcement would seem to be very strong in the North Atlantic area but rational decision-making is not a predictable outcome when considerations of high policy interact with pressures from regional politics. Frustration and conflict in this context could change domestic political alignments and thus affect the littoral states' security-policy orientation, and the future of relations between, for example, metropolitan Denmark, on the one hand, and the Faroe Islands and Greenland on the other. We may see fishing in troubled waters in more senses than one.

Fishing is not the only resource-extracting activity to take place within the EEZ's; drilling for oil and gas on the continental shelves will become increasingly important in the years ahead. So far the major

9

effort has been concentrated in the North Sea, but other areas are likely to prove productive as well, particularly within the Arctic Region. Such activities may give rise to several kinds of conflict: issues related to demarcation and delimitation of shelf areas between adjacent and opposite states will have to be settled, and the status of uninhabited islands determined. Norway and the Soviet Union are currently engaged in difficult negotiations about delimitation of the shelf in the Barents Sea, and a potential conflict is emerging between Iceland and Norway over the dividing line between Iceland and Jan Mayen Island. Previously insignificant islands may become coveted real estate, essential for the establishing of claims to surrounding ocean areas. (Rockall is a case in point). Pressures may be exerted and conflicts arise over the rate of extraction: for instance, Norway has decided on a much more moderate rate of extraction than the United Kingdom,[5] and has at times been severely criticized for it. Recurrent energy crises may cause conflicts between Norway and her major allies over this and, in addition, the oil-producing countries in the North Atlantic area are likely to differ from the non-producers on the question of oil prices. Conflicts over allocation and extraction may arise in oil and gas fields which traverse national boundaries on the continental shelf.

The outer limits of the continental shelf itself may become the subject of dispute. The exploitability criterion of the 1958 continental shelf convention provides no fixed limits, and several states' claims reach as far as the continental margin where the shelf extends beyond 200 miles from the coast. Such claims are, however, by no means universally acknowledged and could give rise to conflicts over rights of access; there are proposals for various kinds of revenue-sharing in the area between the 200-mile limit and the edge of the continental margin for the benefit of developing countries.

Norway confronts a difficult situation over the delimitation of her continental shelf, which is compounded by the special status of the Svalbard archipelago. The Soviet Union, Britain, the United States and France have made reservations with respect to her contention that the shelf surrounding the archipelago is a continuous extension of the Norwegian land mass – and hence belongs to Norway without being subject to any of the limitations in the Svalbard Treaty, which grants equal right of access and establishment to all the signatory powers. Norway also contends that, even if the shelf were to be divided between metropolitan Norway and Svalbard, the Treaty applies only to the resources within the limits of the Svalbard territorial sea. Finally, if the Svalbard Treaty were to apply to a portion of the continental shelf, Norway could reasonably argue that the extent of the Svalbard shelf should be limited to the Svalbard rectangle defined by the Treaty.

(The area between 74° and 81° North and 10° and 35° East.) The international oil companies are likely to be important actors in a dispute over access to the shelf around Svalbard, but the Svalbard mining ordinance would not be an appropriate instrument for managing offshore prospecting and production. The area is strategically sensitive due to the transit routes of the Soviet Northern Fleet and the proximity to possible deployment areas of the Soviet SSBN force.[6] Norway and the Soviet Union are the only countries which maintain permanent settlement on Svalbard (presently in the ratio of 1:2), and Soviet pressure for extensive bilateralism in the area is therefore likely to increase if there is multilateral conflict between Norway and her principal allies over access rights and regulatory powers in the area around Svalbard.

While the dispute about Svalbard is legal in form, it is likely to be resolved as a political matter. Norway cannot, of course, enforce a solution which is unacceptable to the key signatories to the Treaty; furthermore, she would not cherish the prospect of a conflict with the major NATO powers and the Soviet Union at the same time. In pursuing her policy in the area she has to ask herself why other powers should accept the Norwegian position. The potential reasons include the validity of Norway's legal argument, her need of the resources in the area, her ability to manage them, the possibility of rejection producing negative reactions in Norwegian internal politics, and the assessment that the exercise of Norwegian sovereignty in the area, by preventing competition, will improve stability and security. Most of these conclusions are debatable at best. Ultimately, the international interest is to avoid strategically destabilizing spill-over effects from the arena of economic competition. A possible compromise would be to limit the Svalbard regime to the area of the rectangle defined by the Treaty. Norway should in the event of such a compromise probably convene a conference of the parties to the Treaty for purposes of gaining acceptance of a code for petroleum prospecting and production on the continental shelf within the rectangle, as the Svalbard mining code would be unsuitable for these purposes.

Disputes and conflicts may also arise over conflicting ocean usage. The potential conflicts between fishing and oil and gas production are obvious, but conflicts may also arise over artificial islands, offshore production facilities, transportation, scientific research, storage facilities, waste disposal, etc. In many instances these will involve the extent of coastal states' non-resource-related regulatory powers within the EEZ. The right to establish pollution standards in particularly sensitive environments will be of special importance in the Arctic area, and this problem will grow as trans-Arctic transportation of goods and raw

131

materials becomes an economic possibility. (The technology for this is becoming available: for example, the Norwegian-designed 250,000 ton semi-submersible icebreaking tanker).[7] In the Arctic, problems of pollution control, transit and the legal status of ice may generate conflicts between some NATO powers over the Northwest Passage, and between the Soviet Union and western powers over the Northeast Passage.

The North Atlantic States, as maritime nations, have not favoured limitations on the freedom of passage within the EEZ outside the limit of the territorial sea. However, should passage be limited to innocent passage, the viability of the SSBN force concept would be in jeopardy, at least for the Soviet Union – though the littoral states' ability to exclude submerged SSBN passage or patrols even from their territorial seas is not particularly impressive. 'Creeping jurisdiction' could raise security-related issues concerning the installation of fixed sonar arrays and the conduct of scientific research, but such issues are unlikely to produce disputes among the North Atlantic countries, because of their NATO links. There will be no 'territorialization' of important international straits in the North Atlantic area as a result of the extension of the territorial sea limit from 4 to 12 nm.[8]

Offshore operations could in principle provide new targets of attack for purposes of acquisition or denial. It seems unlikely, however, that drilling installations would constitute primary, as opposed to incidental, military targets in a limited or general war in the North Atlantic area. The primary objectives will be related to on-shore installations, (though offshore installations will nonetheless be very vulnerable to military attacks with precision-guided munitions), since any attempt to ensure energy supplies in war would have to concentrate on on-shore stockpiling and distribution systems. We could in principle envisage limited conflicts where military attacks against offshore installations are undertaken for purpose of exerting political pressure, but there are considerable structural disincentives to such acts of piracy. Retaliation in kind would be the obvious military response in many instances, and in others retaliation might involve, for instance, the arrest of an oil tanker belonging to the attacker. Demonstrations of naval power in the vicinity of offshore installlations may nevertheless be a means of 'reminding' the littoral states of the realities of power.

Terrorist attacks constitute another set of contingencies. They do not, however, seem very likely. Offshore installations are not readily accessible and, being robust structures designed for a rather inhospitable environment, are not susceptible to easy destruction. On-shore facilities, like refineries, would appear to offer preferable targets, from the point of view of both accessibility and visibility compared to offshore installations. It has also been suggested that offshore installations may become platforms for clandestine military activities. In principle we could envisage the emplacement of weapons, detection systems and electronic surveillance equipment, but it would be very hard to conceal such activities in a civilian environment, with considerable traffic and turnover of personnel. Alternative basing would seem preferable to ensure both optimum location and security of control.[9]

Norway is specially concerned about the question of military protection for offshore installations. Such protection would seem more feasible in the North Sea than in the Barents Sea, due to the superiority of Soviet naval power in the latter. However, differentiated protection regimes could convey unfortunate implications of differentiated sovereignty. In addition, implementing such measures in the North Sea would influence Soviet expectations about the military activities likely to accompany offshore activities in the Barents Sea, and such expectations would affect the Soviet position in negotiations over division of the shelf area in the north. However, unilateral self-denying ordinances on the part of Norway would tend to strengthen trends in the direction of Soviet dominated bilateralism. The issue is one of political framework rather than defence planning.

The littoral states will acquire the capabilities to inspect installations in order to check safety regulations[10] and see that all is in accordance with the concessions under which they operate. Industrial espionage from naval or commercial vessels may become a special problem of policing. The same is true of the enforcement of safety zones and traffic separation schemes.

PROSPECTS FOR MANAGEMENT AND ARMS CONTROL

The possibility of conflict over access to resources among the North Atlantic countries should not be underestimated. However, conflict is by no means an inevitable outcome of the transition from freedom of access to a regulated regime. It would seem rather important that coastal state regulatory regimes be augmented by regional regulatory authorities, and the two existing fisheries commissions lack authority, being essentially bodies for bargaining over the allocation of catches. All fishing interests should be represented in the regional authorities, but the coastal states in whose EEZ the species in question are found should carry special weight within those authorities.

It is important for the maintenance of regional order that disputes arising out of ocean uses be settled amicably. Mediating panels of acknowledged experts from the region should be established and the states should commit themselves to using them to help resolve disputes. Regional standards for enforcement, and the responsibilities and authority of coastal and flag states, should be agreed on at the regional level for the purpose of maintaining order on the fishing grounds.

Offshore activities could also provide a basis for expanded regional co-operation. The North Sea is in many ways a pilot undertaking, and the model it provides for other areas may be of considerable importance in avoiding and resolving conflicts. A North Sea Operations Authority could be established to provide structure, coherence and efficiency for the extraction efforts. Co-operation could focus on safety regulation, pollution standards, rescue operations, surveillance, dispute settlement and pipelines, and a North Sea constabulatory force and fire brigade could also be envisaged. There is an obvious need for the littoral states to work out such specific rules for the Arctic basin as well, and the Soviet Union would be a necessary partner in such an undertaking. Thus, from the point of view of strengthening the trends towards collaboration across the East-West divide, the Arctic offers the option of an elliptical approach. Such co-operation ought to be structured along functional lines and concentrate on rules of access and transit as well as scientific collaboration.

In the area of arms control maritime forces have received little attention in the post-war period, as opposed to between the wars. For a long time the reason was probably the absence of a comprehensive challenge to American naval preponderance, but the impressive growth of the Soviet Navy, and its shift to forward deployment, have led to renewed interest in naval arms control as a means to contain competition and conflict.[11] We must distinguish between arms-control measures designed to regulate competition in the North Atlantic and the implications of arms control in other areas for the North Atlantic naval environment. In regard to the latter, an agreement on mutual force reductions in Europe, leading to a draw-down of American troops in that area, will increase the importance of the ability to bring them back. Ability to maintain the sea lines of communication across the North Atlantic would thus take on increased importance in the East-West balance. The operational flexibility of aircraft carriers could be reduced if in connection with SALT II a noncircumvention agreement covering forward-based systems were to prohibit them moving within the striking range of Soviet territory. In such circumstances the deterrence inherent in an ability to project power ashore in North Norway would be curtailed; the capability itself would not be affected, of course, but the revised rules of peacetime engagement would influence the political calculus.

With respect to arms-control arrangements in the North Atlantic area we can distinguish between deployment limitations, inventory limitations, sanctuary arrangements and codes of conduct.

The North Atlantic is much too wide and open-ended an area to constitute a possible arena for regional deployment limitations. The north-east Atlantic may offer a more promising possibility, in that it has identifiable geographical boundaries. However, the geo-political configuration militates against any solution limiting naval presence to the littoral states. If the Soviet Union is to be counted as a North-East Atlantic littoral state she would become the unchecked dominant naval power. A regime which excluded her would clearly be unacceptable to the Soviet Union, for whom the north-east Atlantic provides her primary access to the high seas and to patrol areas for her growing SSBN force. We could, of course, envisage a deployment-limitation scheme which gave the US Navy a certain right of presence. Moscow may not be willing to concede equality of presence. Any regime which codified Soviet supremacy would have an adverse political impact on the structure of security arrangements of the littoral states. A deployment-limiting regime based on essential equivalence between the Soviet and American navies in the North-East Atlantic could therefore possibly have a stabilizing effect, provided some simple index to determine volume of presence could be found which would not affect the two superpowers unequally. That will be very hard, due to the asymmetrical composition of the two navies: for example, a simple tonnage restriction might discriminate against the United States, whose ships are bigger than their Soviet counterparts and whose aircraft carriers are her primary means of projecting redressing power ashore in Northern Europe. Maritime surveillance capabilities would be important for controlling adherence to a deployment-limitation agreement.

Deployment limitations would probably have to apply to the number of ships of broadly defined categories. There are two successful precedents for such limitation schemes: the Rush Bagot Treaty of 1817, limiting American and British (later Canadian) naval deployments on the Great Lakes, and the Montreux Convention of 1936, regulating passage through the Turkish Straits and the naval deployments of non-littoral states in the Black Sea. (However, neither agreement regulated deployments in an area which was central to the global balance of power.)

It seems unlikely, though, that the Soviet Union, the United States, Britain or France would find it acceptable for the number of their SSBN patrols in the north-east Atlantic to be constrained by ceilings which are low compared to the present or prospective volume of presence. Nor are they likely to accept ceilings on SSN's escorting or trailing

SSBN's. Moscow, on the other hand, is unlikely to agree to equality of submarine presence.

From the point of view of negotiability a deployment-limitation agreement in the north-east Atlantic does not look very promising. The incentives for circumvention are likely to be rather strong and hence reciprocal suspicions of violations could generate more tension than would exist in the absence of agreement.

An inventory-limitation agreement would be even more problematical. It would have to be general rather than regional in character, on the pattern of the Washington and London agreements of the inter-war period, and the index problem would be very severe, because of the asymmetries in aquired naval forces and the geo-political position of the Soviet Union and the United States. Any agreement of this type to which Moscow would be likely to agree would presumably be based on a parity arrangement, but it is not evident that this would be conductive to stability in the North Atlantic region, where the security structure has been very largely based on the assumption of Western naval preponderance. Adjustments have been made to a more fluid and less calculable situation than that of the 1940s and 1950s, but codified naval parity – however defined – could make Soviet land and air superiority in northern Europe much more of a political burden than it is under present conditions. The US Navy will remain an important instrument in support of American interests outside of Europe, but the American naval presence has in general been concentrated in areas other than the North-East Atlantic, whereas the Soviet Navy exhibits a different pattern. Hence, a general inventory limitation agreement may have an unequal impact on Soviet and American naval presence in the North-East Atlantic. At the same time, it should be recognized that naval limitation agreements constitute an aspect of political agreements about the distribution of political power in certain basins, and that any agreement based on parity, by increasing the Soviet share of regional power, would threaten to upset the current configuration of regional security arrangements in Northern Europe.[2]

It is possible to imagine limited agreements which prohibit or constrain particular capabilities or activities, but it is difficult to envisage consensus on them, given the asymmetries in capabilities, missions and requirements of the United States and the Soviet Union. From the point of view of the littoral states in Northern Europe, Soviet amphibious capabilities are a primary source of threat and American amphibious capabilities a primary source of support. It has also sometimes been suggested that agreed limitations on anti-submarine warfare (ASW) activities would be conductive to stability by eliminating threats to the second-strike missions of the SSBN forces. However, ASW forces are

136

needed not only to hunt or protect SSBN, but also to protect surface combatants and merchant vessels. Constraints on the ability to protect the sea lines of communication would have unequal effects on the security interests of NATO and the Warsaw Pact countries, since the Western alliance is dependent on trans-Atlantic supply lines in an emergency, while the Pact depends primarily on interior, overland supply lines. A limitation on SSN would be possible, provided it included nuclear guided-missile submarines as well: a freeze at, for example, 80 units a side is a possible solution.

Another potential option would be the reciprocal observance of SSBN sanctuaries. However, such agreements would inevitably carry implications for the distribution of political power and influence in the adjacent land area. For example, sanctuary could be envisaged for Soviet D-class SSBN's in the Barents Sea, and this would presumably entail a commitment by Western ASW forces not to enter the sanctuary zones, and, in particular, to exclude them from their maritime surveillance envelopes. But such a change of pattern could affect the geopolitical viability of continued Norwegian alignment and involvement in the NATO system of maritime surveillance. Pressures for expanding 'sanitization' to include the littoral areas as well could increase with expanded civilian presence in the SSBN sanctuary areas in connection with oil and gas production on the continental shelf. The sanctuary could thus become the basis for expanded claims to preferential access to resources and modification of the dividing lines on the continental shelf. Finally, it is not evident that concentrating SSBN's in limited sanctuaries would be conducive to stability (due to target concentration) or to deterrence (due to attack route concentration).

A more promising avenue would be the gradual contractual and tacit crystallization of codes of conduct and reciprocal restraint. Such codes could involve agreement not to establish SSBN or SSN bases in the vicinity of the home port areas of the adversary's SSBN forces. In concrete terms, it would involve a Soviet commitment not to establish such bases in Cuba and a reciprocal American commitment not to do so in Norway. There could also be agreement not to interfere with maritime surveillance activities, in recognition of the stabilizing effect of the ability to ascertain the absence of feared threats. The United States and the Soviet Union concluded an agreement on the prevention of incidents involving warships at sea on May 25, 1972, and it is possible to envisage an expanded regime of reciprocal restraint involving a restraint on the duration of naval manoeuvres. The provisions of the Final Act of the Conference on Security and Co-operation in Europe covering advance announcement of military manoeuvres on land could be expanded to include naval manoeuvres as well. Further-

137

more, all states could agree not to carry out naval exercises or major naval movements closer than 50–100 nautical miles from the coasts of non-participant states. Similarly, a commitment not to stage amphibious landing exercises closer than 50–100 km from the territories of adjacent states would be another possible measure for reducing the threat of naval pressure in the political environment of the littoral states.

There is little prospect of complete removal of this threat, but some amelioration would be an important contribution towards expanding the spatial coverage of the emerging détente to embrace the North Atlantic. Functional expansion to include naval forces, in addition to land and air forces, would emphasize the interdependent nature of the security structure between East and West. A system of reciprocal restraints covering naval activities would presumably also be conducive to the construction of a maritime order embracing all the uses of the oceans in the North Atlantic area.

GEORGE R. LINDSEY

The Future of Anti-Submarine Warfare and its Impact on Naval Activities in the North Atlantic and Arctic Regions

THE NAVAL PROBLEMS OF THE NORTH ATLANTIC AND ITS ARCTIC FLANK

For four centuries the North Atlantic has borne the ocean traffic between Europe and America. Looking to the future, the Arctic Ocean may well become a second Mediterranean – a long sea between busy and important continents, and likely some day to offer communication between them. The air routes – now more important than the sea for passengers, though certainly not for goods – span much of the Arctic as well as the North Atlantic. Any strategic studies of communications between Europe and North America must focus on the North Atlantic, and may or may not encounter special problems on its Arctic flank.

Until the appearance of the strike carrier and the missile-firing submarine, the naval strategy in the North Atlantic had as its main concerns the defence of the sea routes by the powers working to maintain communication, and the attack by those wishing to enforce a blockade; expressed in other terms, these are the roles of sea control and sea denial. The blockaders could attack merchant ships directly, or attempt to destroy the naval forces protecting them. The defenders could attempt to destroy the attackers wherever they could find them, or they could deploy their forces for close escort of the merchant ships. When the Allied blockaders dominated the North Sea it was possible to slip occasional surface raiders from Germany into the Atlantic and beyond, but the main fleet would have to fight its way out and back. Only submarines could hope to make repeated forays from German ports against the trade routes without having to face serious threats in transit. But when the German blockaders could operate from the Bay of Biscay, they enjoyed direct access to the North Atlantic trade routes.

Direct projection of sea power ashore, possible for centuries through the mechanisms of naval bombardment and amphibious landings, was exerted by the Allies in the second war in the Mediterranean, during

139

the Normandy landings and at many points in the Pacific theatre, where operations benefited enormously from strike carriers. An important aspect of naval strategy in the North Atlantic and its Arctic flank today is the operation of NATO carrier strike forces off the coast of Europe and of Warsaw Pact aircraft, submarines and surface ships.

The ultimate development for projection of sea power ashore has been the ballistic-missile-firing submarine. In terms of fire-power, range and invulnerability, it considerably exceeds the strategic capabilities of the carrier; where strategic nuclear deterrence is involved, the strike carrier has been superseded by the SSBN. But carrier forces can perform many tactical roles, especially in circumstances involving close support of troops ashore, and where nuclear weapons are excluded. It is probable that the future roles of aircraft carriers in the North Atlantic will be more tactical than strategic – support of conventional operations ashore, and operations to secure sea control against opposing surface ships, submarines and aircraft.

World War II saw action by major surface ships in the North Atlantic, including the sinking of the *Hood* and *Bismarck* at sea by gunfire and torpedoes from aircraft and surface ships, and the *Courageous* by a U-boat. Many other German warships were badly damaged in port by air attack, and the *Royal Oak* was sunk in harbour by a U-boat. German surface raiders of all sizes sank about 6 per cent of the total Allied merchant tonnage destroyed, about half of this in the North Atlantic. When in June 1941 heavy surface escort aided by intermittent air escort was provided right across the North Atlantic there were no more sinkings by surface raiders in that area. The withdrawal of Allied forces across the English Channel from Dunkirk in 1940 required a large number of vessels of all sizes (more than 800), including many smaller naval units; 9 destroyers and numerous smaller craft were lost, mainly due to air attack. Many large troops convoys from Britain to the Mediterranean or navigating the Cape of Good Hope had to traverse part of the North Atlantic en route. The Normandy landings of 1944 were conducted with very heavy naval support, including 7 battleships, 2 monitors, 23 cruisers and 105 destroyers and of the 9 destroyers lost, 6 were sunk by mines but none by submarines.

Nevertheless, the naval operation that dominated all others in this part of the world (in importance and in number of forces involved) was the maintenance of merchant shipping between the United States and Britain in the face of attack by German aircraft and submarines. Of the huge total of 21.6 million tons of Allied and neutral merchant shipping lost by enemy action in World War II, 55 per cent went down in the North Atlantic and Arctic, with another 18 per cent lost in British coastal waters, the North Sea and the Baltic; 14.7 million tons (about 68 per cent) were sunk by German, Italian and Japanese submarines, and 2.9 million tons (about 13 per cent) by aircraft.

As a comparison, about 88 per cent of the 12.7 million tons of Allied and neutral merchant shipping lost by enemy action in World War I was sunk by U-boats, a large proportion in British waters and the Western approaches. The total of 849,000 tons sunk by enemy action in April 1917 exceeded the maximum for any month in the Second World War (834,000 tons in June 1942, of which nearly 70 per cent was in the Atlantic and 84 per cent due to the submarines). The Germans calculated that a sustained loss of 700,000 tons per month would lead to Allied defeat in World War II. They were able to exceed 650,000 tons in nine of the months between 1941 and 1943, including a score of 728,000 tons for U-boats alone in November 1942. In the ten months between February and November of 1942 they sank over 7 million tons, 80 per cent by U-boats and about 70 per cent in the North Atlantic.

The defence against the U-boats had three aspects. The most important was the organization of merchant ships into convoys escorted by ships and aircraft. The others were offensives against the U-boats while in transit between their bases and their operating areas, and aerial bombing of U-boats in port or under construction.

Anti-submarine convoys were first introduced in the summer of 1917, and they produced an immediate and permanent reduction in losses. They were used across the North Atlantic from the beginning of World War II. However, between December 1941, when the United States entered the war, and May 1942, shipping off the American coast sailed independently and lost 1.2 million tons to a force of no more than a dozen U-boats operating in the Western Atlantic. Introduction of convoys brought the situation under control.

The main convoy routes were between New York or Halifax and Britain, New York and Gibraltar, Britain and Gibraltar, Britain and South Africa. (See the top two maps of Figure I). Routes were changed to take advantage of friendly land-based air cover, as this became available in Newfoundland, Iceland, Greenland and the Azores, and were kept away from enemy air bases in France. The U-boats also changed their operating areas, depending on circumstances; many were withdrawn from the North Atlantic when operations were pending in Norway in March 1940, when there was particularly severe winter weather the same year, when there were rich opportunities in the Western Atlantic and Caribbean in 1942, when the landings in North Africa were made in 1942, and when the cross-Channel landings were expected in 1944. In the North Atlantic the areas of U-boat concen-

141

FIGURE I

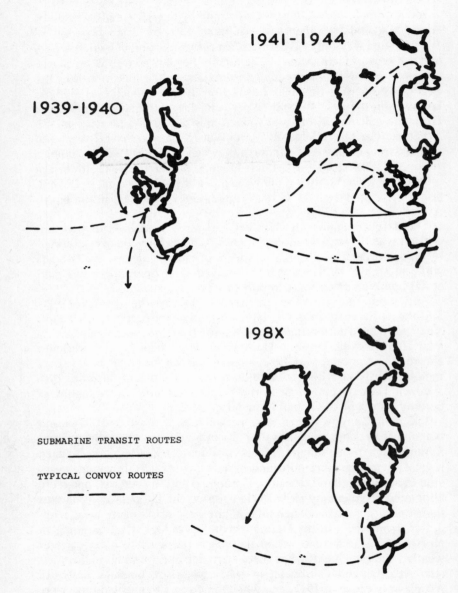

1941-1944

1939-1940

198X

SUBMARINE TRANSIT ROUTES

TYPICAL CONVOY ROUTES

tration changed as land-based air cover was extended by new bases and the appearance of longer-range aircraft.

One measure of success for the anti-submarine forces is the number of submarines sunk. In this there is a remarkable contrast between the first thirty-four months of the war, in which the Allies sank 78 German U-boats, and the second thirty-four months, in which no less than 707 were destroyed. The merchant tonnage sunk by submarines in each half of the war was approximately equal (7.9 million and 6.8 million) so that the exchange ratio improved by a factor of about ten. The explanations for the improvement are many, and include the use of air and naval bases in Iceland, Greenland, and the Azores, the building of large numbers of long-range maritime aircraft and improved escort ships, the introduction of escort carriers, the development of new anti-submarine weapons, such as the airborne homing torpedo, ahead-thrown depth charges, a series of advances in electronics, such as shipborne and airborne metric and then centimetric radar, shipborne high frequency direction-finders, sonobuoys to allow aircraft to maintain contact with submerged submarines, the Leigh light to enable aircraft to illuminate a submarine caught on the surface at night, and the magnetic anomaly detector, permitting aircraft to sense the presence of submerged metallic objects. Also very important was the development of the proper tactics by which all of these devices could be fully and effectively exploited. Two-thirds of the U-boat sinkings were achieved by ships and shore-based aircraft, in almost exactly equal proportions, often working in combination.

Improvement of tactics and equipment was not confined to the anti-submarine forces. However, the technical developments for the U-boats did not receive high priority in Germany, and many of the most significant improvements came too late to have much result. Acoustic homing torpedoes were effective against escort ships. The Snorkel afforded very important protection to those U-boats which finally acquired it during the last year of the war. The greatly enhanced performance of the new Type XXIII and XXI U-boats was hardly put to the test, since neither were used in operations until 1945. German appreciation of the effectiveness of airborne radar was very slow, although they did eventually fit the U-boats with satisfactory warning receivers.

In opposition, the Allies exerted a blockade against Axis sea communications. The Germans were able to maintain control of the Baltic and the routes to Norway and Sweden, but were soon cut off from any substantial merchant traffic to the Atlantic. This blockade depended more on surface ships, aircraft and mines, than on submarines.

The employment of anti-submarine forces to attack submarines in transit between their bases and their operating areas, or to hunt them in the open ocean, had its main opportunity when the German U-boats based on the west coast of France had to transit the Bay of Biscay to reach the North Atlantic. During a period of three months in 1943, when they elected to move on the surface in groups, and to fight air attack by anti-aircraft fire, the results of the transit patrols were very satisfactory in terms of U-boats sunk (28 representing 10 per cent of the number attempting the transit) and in flying hours per kill. However, during the much longer period when the U-boats travelled singly and submerged by day, aircraft killed over four times as many U-boats per flying hour, when flying in support of convoys, than on the Biscay transit patrol. Moreover, the presence of aircraft over convoys often prevented the U-boats from attempting an attack on the ships. The overall statistics indicate that aircraft and ships were better employed in close escort and support of convoys than in the attempt to catch submarines in transit or to hunt them in the open sea.

Another form of barrier to the transit of submarines is the placing of mine fields across their expected paths. Mines destroyed more German U-boats than any other cause in World War I (42 out of 199). The great North Sea mine barrier, with over 71,000 mines extending from the Orkneys to Norway, accounted for 3 U-boats only. Most sinkings occurred close to the French coast. In World War II mines were laid between the Orkneys and Iceland, along the coasts of the North Sea and up the Norwegian coast, in the Kattegat, along the French coast, and at the entrance to the Irish Sea; 25 U-boats were destroyed, mostly near the French and German coasts. The Iceland–Faroes–Britain minefield, which required an enormous effort to sow, only sank one U-boat, and did not prevent their passage through the GIUK gap.

The third form of anti-submarine operation was direct bombing of the bases and the construction yards. This activity absorbed a very substantial proportion of the wartime bomber effort, and employed many four-engined aircraft, which could have been used instead for anti-submarine patrol. In terms of aircraft lost per U-boat destroyed, the ratio favoured the maritime role by a large factor, and to this one should add the further very important contribution of the maritime aircraft in hindering the U-boat attacks on the merchant ships.

It should be added that submarines had roles other than attack on shipping by torpedoes or gunfire; they laid many mines, provided reconnaissance, delivered supplies and agents, as well as rescued crews of aircraft that had been shot down.

WORLD WAR II NAVAL OPERATIONS IN THE ARCTIC

The most important military operations in Arctic waters during World War II were the passages of merchant convoys from Scotland and Iceland into Murmansk and the White Sea. Heavy losses were inflicted on them by German aircraft, U-boats and surface ships able to take advantage of bases in Norway, but the Germans also suffered considerable losses in aircraft and submarines. Heavy German warships anchored in Norwegian fjords posed a deadly threat to the Murmansk convoys, only partially offset by the protection of British battleships operating out of Iceland and the Orkneys. Eventually *Scharnhorst* was sunk at sea off the North Cape, by surface ships, while *Tirpitz* was repeatedly damaged while at anchor and finally destroyed by a series of attacks by midget submarines and carrier- and land-based bombers (including some British bombers operating out of Russian bases). The Royal Navy based minesweepers in Murmansk to counter German mines laid in the approaches. The success of the German operations against the Arctic convoys depended greatly on aerial reconnaissance, which was highly effective in the long summer days but far less so in the winter Arctic nights. In the summers of 1943 and 1944, as visibility improved, increased losses coupled with limitations on northward routing occasioned by the seasonal movement of the Arctic icepack meant that the operation of convoys had to be suspended.

The Norwegian bases would also have been of major importance to U-boats operating into the Atlantic, had it not been for the German occupation of France, which allowed them even better situated bases on the Bay of Biscay. The rate of loss of merchant ships in the convoys to North Russia was 5.7 per cent, eight times the overall rate of 0.7 per cent for all ocean convoys. Of 85 merchant ships sunk, 39 were dispatched by submarine and 37 by aircraft.

There were other naval operations in Arctic waters at the time of the German invasion of Norway in 1940. Narvik was captured by the Germans, but the escorting destroyers were trapped and sunk. The Allies sent an expedition to recapture Narvik, but their subsequent withdrawal provoked a naval battle involving heavy units on both sides and the loss of the British carrier *Glorious*. In the Norwegian campaign altogether, the Germans lost 3 cruisers, 10 destroyers and 4 submarines; the Allies lost 1 carrier, 2 cruisers, 9 destroyers and 5 submarines. Most sinkings were caused by aircraft or by gunfire. Many German U-boat attacks on Allied warships were foiled by malfunctioning torpedoes.

In addition to the establishment of bases in Iceland and Greenland, the Allies occupied the Faroes and had weather stations in scattered locations, including Svalbard. The Germans installed weather stations in Eastern Greenland, Svalbard, Novaya Zemlya, and Franz Josef Land.

The intensity of the northern submarine war is demonstrated by the fact that 62 U-boats were sunk in the Arctic Ocean or Norwegian Sea, compared with 356 German, 13 Italian and one Japanese submarine sunk in the North Atlantic, and 134 German and 1 Italian in the North Sea.

LESSONS OF WORLD WAR II APPLICABLE TO FUTURE ANTI-SUBMARINE WARFARE IN THE NORTH ATLANTIC

In order to assess the future of anti-submarine warfare in the North Atlantic, we must learn the appropriate lessons from history. But it is important to recognize the very substantial changes that have occurred since 1945, both in the line-up of nations likely to be contestants and in the technological developments that have occurred in the past thirty years. It should be possible, surely, to contemplate the next war without either ignoring the last one or assuming that it will be repeated without change.

Before June 1940 the continental power had very restricted access to the North Atlantic: the Frisian coast of Germany on the North Sea, and the Baltic exits past neutral Denmark, Sweden and Norway. From the North Sea it was impossible to transit the English Channel with hostile powers on both sides. The only exit from the North Sea was into the Norwegian Sea, between Norway and the Shetlands. To pass into the North Atlantic it was necessary to traverse the GIUK gap, which was patrolled by Allied naval and armed merchant cruisers, but proved penetrable by surface ships as well as submarines.

After June 1940 the continental power held the French, Belgian, Dutch, Danish and Norwegian coasts. Bases on the Bay of Biscay gave direct access to the North Atlantic, and possession of the Norwegian coast made entry into the Norwegian Sea easy, as well as ensuring the supply of iron ore from Narvik to Germany. At this time the Allies set up bases in Iceland and installed the Iceland-Orkneys mine barrage, so as to make the GIUK gap less penetrable, and they routed the Atlantic convoys to the north of Ireland (see Figure I).

After June 1941 the Soviet Union joined the Atlantic Allies, and provided bases in the Russian Arctic, but merchant convoys there consequently had to pass the now hostile Norwegian coast.

The final naval geo-political change came in the summer of 1944, when the Germans lost their bases on the Bay of Biscay and transferred their submarines to Norway.

146

What comparison can be made for a contest between NATO and the Warsaw Pact? With present territory intact, the Pact is in a position not unlike that of Germany in 1939. It has access to the Norwegian Sea, but now from the bases in the Kola Peninsula and the White Sea. The Baltic exits are now flanked by two NATO opponents and one neutral, and there is no equivalent to the German bases on the North Sea. But the GIUK gap is better defended than in 1939, with Iceland and Greenland in the NATO camp.

One could attempt to compare the passage of Allied shipping from Iceland to Murmansk and Archangel, past the hostile Norwegian coast, with the passage of Soviet warships from Murmansk to the GUIK gap. If this passage can be contested from an unfriendly north Norwegian coast and the Arctic islands of Iceland and Greenland, then the passage of surface ships could be made very difficult. It is easy to see why the Warsaw Pact is anxious to bring about at least the neutralization of all of the Arctic islands, including the Faroes, Svalbard, Bear, and Jan Mayen, as well as Iceland and Greenland.

As a next step, and remembering that the strongest element of the Soviet Navy is based on its North-West Arctic coast, it would appear fairly clear that the preeminent objectives of Soviet naval policy would be to neutralize or otherwise secure the Norwegian coast, to guarantee unimpeded access to the Norwegian Sea, and thereafter to take steps to ensure that they were able to transit the GIUK gap with surface ships as well as submarines.

Other lessons learned during World War II, but possibly superannuated by subsequent technological developments, included one of overriding importance reinforced by the experiences of World War I. This was that collection of merchant ships into convoys and provision of close escort ships and aircraft was absolutely essential, with the exception of fast units which could take advantage of speed and evasive routing and tactics. Related to this was the lesson that anti-submarine ships and aircraft were usually better employed in the close escort or support of merchant convoys than in contesting submarine transits or in attempting to sweep the open ocean. There were, however, certain special circumstances in which submarine transits could be opposed quite effectively.

The final anti-submarine lesson of World War II, that large aircraft can oppose submarines more effectively by maritime operations than by attempting to bomb them in their bases, is probably not relevant to the conditions of the 1970s.

POST-WAR DEVELOPMENT OF ANTI-SUBMARINE TECHNOLOGY

Anti-submarine technology has made notable progress since 1945: sonar has been markedly improved and is now installed in fixed systems on the seabed, attached to the hulls of ships,towed below ships, suspended below hovering helicopters and below floating sonobuoys with a radio link to aircraft, ships or helicopters.

Anti-submarine ships can now project their torpedoes or depth charges to a considerable distance using mortars, rockets, cruise missiles or helicopters. The larger helicopters also carry dipping sonar as well as torpedoes, and some can sow sonobuoys as well as receive information from them. Very powerful active sonars have been mounted on the hulls of surface ships, but it is likely that further progress will take the form of equipment towed behind and below the ship, where it is separated from the ship's noise and able to operate at a depth better suited for long-range propagation of sound.

The large anti-submarine aircraft carrier is being replaced by the multi-purpose carrier and smaller ships known as Through-deck Cruisers or Sea-control Ships able to carry anti-submarine (AS) helicopters or other vertical take-off aircraft.

As a general trend, it appears that the anti-submarine submarine has taken a leading place in the roster of AS weapon systems. The AS aircraft has kept its place, but the AS escort ship has dropped back, unless it is equipped with a helicopter. Since the range of submarine-launched anti-shipping missiles is likely to exceed the range of detection of the submarine by an escort ship, the only role left for the escort ship with no helicopter may be one of anti-aircraft protection; that is, to defend the convoy against aircraft and to attempt to destroy cruise missiles while they are in flight toward the ships.

Anti-submarine mines are now more sophisticated, and can even take the form of a capsule able to launch a torpedo from information received from its sensors.

POST-WAR DEVELOPMENT OF SUBMARINE TECHNOLOGY

As has been mentioned already, important improvements to submarines appeared very late in World War II. The momentum continued afterwards with the Snorkel becoming standard equipment for all diesel-powered submarines, and German improvements in underwater performance. Passive sonar has become much more effective. Anti-shipping torpedoes have been given greater range and can now home with self-contained active or passive guidance, or by signals sent by wire from the launching submarine. Anti-shipping cruise missiles have been fitted (especially by the Soviet Union), allowing the engagement of surface targets at very long range.

But two quite different technological breakthroughs have given submarines capabilities far beyond those of 1945; these are nuclear propulsion and the submarine-launched ballistic missile (SLBM). Nuclear propulsion allows a submarine to have unlimited submerged endurance, with no requirement to project anything above the surface for months at a time. And the SLBM puts the submarine into the strategic role of projection of power ashore, which is quite separate from its long-established anti-shipping function.

Nuclear propulsion contributes a great deal to the effectiveness of the submarine in the anti-shipping role. In transit it need offer no clue above the surface for searching aircraft, such as a Snorkel mast or diesel exhaust. Now the submarine can move faster than the convoy without having to come to the surface. If engaged by anti-submarine forces, it will not be forced to come up to recharge exhausted batteries. A submarine equipped with both nuclear propulsion and cruise missiles is a very much more formidable anti-shipping weapon than the World War II diesel-powered U-boat, armed with torpedoes and a gun. However, it should be remembered that more than two-thirds of the world's submarines are diesel-powered, including 173 conventional torpedo attack submarines in the Warsaw Pact countries, so that it will be some years before the most effective types are also the most numerous.

Another role which has developed for the modern submarine is the destruction of hostile submarines. In fact about 65 submarines were destroyed by other submarines in the course of World War II; in almost all cases the victim was caught while on the surface. But submarines are now being designed specifically for the anti-submarine role, with very quiet nuclear propulsion, extra sensitive passive sonar, and special anti-submarine torpedoes. A nuclear depth bomb can be launched from a torpedo tube and then projected through the air by rocket to the vicinity of its target before it re-enters the water.

The strategic nuclear-powered ballistic missile submarine (SSBN) represents the greatest development in naval technological history. SLBM can be launched while the submarine is submerged, or they can be sent in a salvo with only a brief interval between launchings. The accuracy may not quite match the best that can be achieved with land-based missiles, but it is more than adequate to deal with almost any strategic target of known position, other than a hardened missile silo. Multiple warheads can be mounted, as well as all the penetration aids available to ICBM (inter-continental ballistic missiles).

THE SSBN AND STRATEGIC ANTI-SUBMARINE DEFENCE IN THE NORTH ATLANTIC AND ARCTIC

The strategic missile-firing submarine represents a complete departure from previous submarines and it must be considered, along with the ICBM and the strategic bomber, as an element in the strategic striking-power and the nuclear deterrent. Though this is quite separate from the older role of economic blockade, this does not mean that the latter, and the role of the attack submarine, have disappeared. While there are only 124 ballistic missile submarines, there are 553 ocean-going attack submarines at sea today, excluding the 106 nuclear-powered submarines (SSN) which can be used against shipping or other submarines. The strategic and tactical roles of the submarine are absolutely different and must be considered separately. Unfortunately, for purposes of discussion and analysis, most of the anti-submarine measures that can be taken against strategic and tactical (anti-shipping) submarines are the same, so that the two roles tend to become confused when considering the counter-measures. To add to the confusion, the strategic missile submarines are also equipped with torpedoes.

The primary role of the SSBN is to threaten the cities of the opposing country with nuclear retaliation in the event that its homeland is attacked. Another role could be in support of a first strike against opposing strategic forces, probably with targets such as airbases, communication centres and seaports, for which pinpoint accuracy is not essential but minimum warning-time is likely to improve the strategic effectiveness of the attack.

For the first purpose, the range at which the missiles are fired is not important. In 1975 *Polaris* and *Poseidon* could reach 2,500 nm, SS–N–6 1,500 nm, and SS–N–8 4,200 nm. By 1977 *Trident* I will have a range of 4,000 nm and some time after 1980, *Trident* II should reach 6,000 nm. So, to contribute to the nuclear deterrent, SSBN can patrol thousands of miles from their targets. Moreover, as they are built to remain undetected, they can make their approach in acoustically unfavourable areas. In terms of North Atlantic geography, Y-class submarines with 1,500 nm missiles need to operate only in the western half of the Atlantic and D-class with 4,200 nm missiles need not cross the GIUK gap. *Polaris* and *Poseidon* submarines with missile ranges of 2,500 nm do not need to pass to the north of the GIUK gap. SSBN can stay so far away from potentially hostile shores that there is little possibility of detecting them unless the whole ocean can be insonified, or unless special anti-submarine submarines can pick them up as they leave port and trail them continuously afterwards. Failing these developments, SSBN remain confirmed as the most invulnerable component of the strategic deterrent, rather like undiscoverable and untargetable ICBM.

However, SSBN will always retain their potential for the first-strike role. To be most effective for this purpose they will need to come much closer to their targets than the many thousands of miles permitted by the maximum range of the latest missiles. Any missile with sufficient propulsive power to attain a range of several thousand miles could be fired on a low-angle trajectory at much shorter range, greatly shortening the time of flight and making radar detection more difficult. 'Time sensitive' targets such as airfields or command centres could be caught with their aircraft on the ground and key personnel not in position. But knowledge of the approach of SSBN with long-range missiles close to the coast would give warning of the preparation for a first strike. Conversely, reliable knowledge that no such submarine had come close to the coast would confirm a situation of minimum threat and maximum stability, and reduce the likelihood of nervous over-reaction in a crisis.

Unless an SSBN has been tracked with great accuracy, and its position well established at the moment when missiles are launched, there is virtually no possibility of preventing the release of the full salvo. Consequently, the measures against the SSBN are likely to be confined to intelligence-gathering to gain general information on its movements, supplemented by a more concentrated effort to fix its position and track its movements if it should come within a much closer range of counter-force targets than would be justified by its second-strike deterrent role. In terms of the North Atlantic and Arctic Oceans, this means that NATO will become specially alert for Soviet SSBN coming south of the GIUK gap, while the Russians could pay very close attention to NATO SSBN coming into the Norwegian Sea or beyond. Thus, the GUIK gap is likely to become a very important line of demarcation for strategic anti-submarine defence.

If a land war should break out in Europe there would be an immediate need to rush military supplies across the Atlantic. Should hostilities continue for weeks, months or years, there would possibly be another 'Battle of the Atlantic', in which NATO's survival would depend on the arrival of sufficient tonnage from America, with the Warsaw Pact attempting to inflict such attrition as to prevent the delivery of adequate supplies.

The strategic use of nuclear weapons on any but the most restricted scale would probably bring such a sea struggle to an end. However, with the undoubted capability of both the United States and the Soviet Union to destroy each other, it does not seem implausible to suppose that this ultimate and disastrous step would be avoided, in spite of the great provocation of conventional war on land and sea. It is even conceivable that tactical nuclear weapons could be used, on land and/or at sea, without subsequent escalation to the strategic use of nuclear weapons.

Under such circumstances, the strategic submarines of both sides would take the greatest precautions to remain as invulnerable as possible, while positioning themselves so as to be able to strike strategic targets at short notice in case the ultimate stage of escalation were reached. There would be little advantage in coming close to the opposing coast, since in a state of war all precautions such as dispersal of aircraft and full manning of control centres would have been taken. The SSBN would seek patrol stations far from areas in which anti-submarine forces were likely to be operating, and they would themselves adopt modes of operation that would make them least likely to be detected. Moreover, the anti-submarine forces of each side would have little reason to hunt the opponent's strategic submarines unless they had decided to deliver an all-out nuclear strike. Operations to attack the SSBN could well be interpreted as the signal that a strike was pending, and hence trigger the final disastrous exchange.

The next section will discuss protection of shipping, the central problem of tactical anti-submarine defence. It will be assumed that the strategic submarines would take considerable pains to keep themselves far removed from the contest for maintenance of the sea routes, and that few, if any, would attempt to transit the GIUK gap. Any doing so would obviously have to take their chance, as if they were attack submarines, and their loss in or beyond the GIUK gap should not be interpreted as an action designed to eliminate strategic submarines. As for nuclear attack submarines, they are effective in three different roles: anti-SSBN, which is the strategic role; anti-surface ship (including attacks on warships and military task forces as well as merchant ships); and the tactical anti-submarine role. In the discussion of the defence of shipping we will suppose that some SSN would be used to attack shipping and others to attack anti-shipping submarines.

DEFENCE OF SHIPPING IN THE NORTH ATLANTIC

Whenever the question of defence of shipping in the North Atlantic arises, the objection is raised that this implies a long war, whereas most NATO planning is concentrated on a short land/air war in Europe. The fact is, of course, that nobody knows whether there will be a long or a short war, but we wish to conduct ourselves in such a way as to bring about a third and very much preferred outcome – that of no war at all. There is value in planning for a long war, including de-

Reminder   Prof Hartmann   21 Nov

0830 - Your seminar.

1830 - Quindecim

fence of the sea lines of communication, whether or not we assign a high probability to its actual occurrence, just as we plan for a strategic nuclear exchange without necessarily believing that it is very likely to occur. In fact, by being seen to be able to carry out these operations successfully, we are making it far less probable that we will ever have to carry them out. This is an extension of the concept of deterrence.

If another 'Battle of the Atlantic' should be fought between NATO and the Warsaw Pact, each retaining its present territories, all the vital transatlantic shipping would be moving on routes well to the south of the GIUK gap. Indeed, if NATO could seal the exits from the Baltic and the Mediterranean, transatlantic convoys could then make their distance from the Kola bases as great as possible by selecting routes passing south of the Azores. At the same time this would increase the distance which would need to be flown by anti-shipping aircraft based in Warsaw Pact territory. Bases in the Azores, Portugal and France would be extremely valuable for the defensive air coverage of the eastern Atlantic.

The changes in submarine and anti-submarine technology, already described, appear to strengthen the capability of the anti-shipping submarine, in spite of the defences that can be provided in close escort or support of merchant convoys. Increased speed and improved sonar give the submarine a better chance of getting into a favourable position to attack. And, reinforcing this advantage, the modern maritime reconnaissance aircraft with its excellent radar and communications, long endurance and high speed, together with reconnaissance satellites, make the presence of a sizeable group of surface ships unlikely to remain a secret for many hours after its assembly. Moreover, once a submarine equipped with cruise missiles makes contact, it can complete its attack without having to come within range of the convoy's close-escort ships.

With the surface ships out-ranged by the submarine, and the aircraft deprived of the opportunity to detect conning towers, or even Snorkel tubes in the case of nuclear-powered submarines, the anti-submarine defence of convoys is going to be very difficult. The most promising anti-submarine vehicle is the nuclear-powered submarine. But it cannot exploit its potential to the best advantage when obliged to move fast for a long time, which would be necessary should it attempt to provide anti-submarine escort to a convoy.

It has been suggested that under these new circumstances shipping should sail independently. This would present the submarines with many scattered targets, allow each ship to move at its maximum speed, and expedite the process of loading and unloading. However, the anti-shipping threat is not confined to submarines, and defence of shipping against air attack can be very much more effective if the targets and their anti-aircraft protection can be concentrated. Moreover, ships equipped with anti-aircraft missiles or guns can use them against cruise missiles in flight, as can land- or sea-based fighter-type aircraft provided for air defence. The safety of the fast passenger ships ('monster liners') which sailed alone during World War II was due to the fact that their speed exceeded that of the submarines. Speed is unlikely to afford much protection against modern submarines with cruise missiles or against aircraft.

Thus, it appears that NATO's transatlantic merchant shipping should be sent in convoys, and these will be routed far to the south of Greenland and Iceland.

However, if anti-submarine defence of the convoys proves to be very difficult, and cannot exploit the anti-submarine capabilities of the submarine, the alternative of combatting the submarines in transit, instead of near their targets, becomes more attractive. The obvious place to establish a transit barrier is across the GIUK gap. This is partially blocked by land, and the flanking routes to the west of Greenland or through the English Channel should be comparatively easy to defend. There is shallow water between Iceland, the Faroes and Scotland, which offers an opportunity for minefields employing improved modern techniques. With air bases in Greenland, Iceland and Scotland, NATO is well placed to contest the passage of submarines through the gap.

The nuclear-propelled submarine should be an extremely effective component of an anti-submarine barrier, especially if efficient co-operation with aircraft is also achieved. This is not a simple matter, and requires good communications equipment as well as practical and well-understood procedures. In the western half of the Denmark Strait, which is covered by ice for a good part of the year, the barrier submarines could not count on the co-operation of either aircraft or surface ships.

RELATED MARITIME CONSIDERATIONS
The two oceans discussed in this paper have very different significance for the NATO nations and for the Soviet Union. The North Atlantic Treaty is well-named: the participants must be able to use the North Atlantic Ocean in peace and in war if the Alliance is to live. In any serious confrontation between NATO and the Warsaw Pact countries the whole basis of NATO's strength would be dissolved if there were serious doubts as to the ability to maintain trade and the transfer of materiel of war across the North Atlantic. Without the North Atlantic there can be no significant North Atlantic Treaty Organization. The

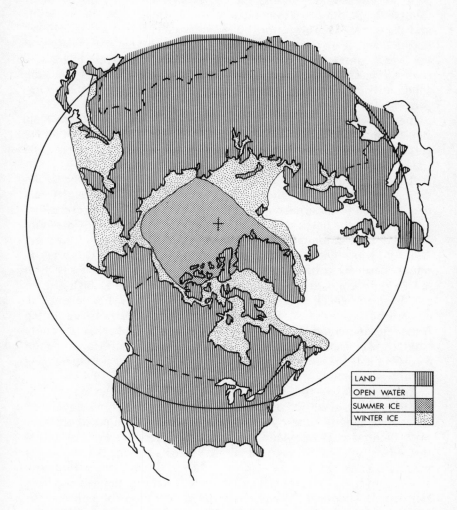

LAND	▦
OPEN WATER	
SUMMER ICE	⣿
WINTER ICE	⣿

Warsaw Pact nations also use the North Atlantic for peacetime trade, and for fishing. But they could exist without it, and, in the event of a clash, their interior continental lines would be much to their advantage.

In the case of the Arctic Ocean, the situation is very different; most of the Arctic Ocean is permanently covered by ice (see Figure II), and almost all of it, except the Norwegian Sea, is frozen during the winter. The area open to navigation in the summer forms an incomplete

155

peripheral rim, with a wide southward opening into the Norwegian Sea, the Greenland Sea, and the GIUK gap. The rim of open water follows the Soviet coast round to a narrow opening at the Bering Strait, and then extends a bit further, past Alaska and the Mackenzie delta, before freezing up against the islands of the Canadian archipelago and the north coast of Greenland. In summer, only 4 per cent of the ice-free Arctic Ocean lies north of Alaska, Canada and Greenland; the other 96 per cent lies above Eurasia and the North Atlantic approaches to Northern Norway and the Murmansk coast. During the winter the Norwegian Sea is free of sea ice, as is the Soviet Kola Peninsula to a point east of Murmansk. But the White Sea and all the rest of the northern shore of European and Asian Russia are impassable to surface traffic, as is the Arctic coast of North America.

It is evident that the Arctic Ocean is far more important to the Soviet Union than it is to any other country in the world. The Northern Sea route is used for internal communication with the Soviet Arctic ports, and between the north-west and the Pacific. Effective icebreaker support keeps the communications open for up to five months of the year. But, for both economic and military purposes, the Soviet Union needs unimpeded year-round access to the world's oceans. From her ports in the Baltic she must traverse the narrow, easily closed Danish Straits; exit from the Black Sea is easily controlled at the Dardanelles. The eastern ports on the Sea of Japan and the Sea of Okhotsk are often frozen in, and could be cut off by the Japanese island chain. It is only from the Murmansk coast that the Soviet Union can sail clear of ice and foreign coasts the year round. This is why Murmansk is now the largest naval base in the world.

As a preliminary to a maritime confrontation, or as an early initiative after one had started, an obvious priority for Soviet strategy would be neutralization of the NATO air and naval bases in Northern Norway, Iceland and Greenland. Actual occupation would not be necessary if NATO's operation could be prevented by diplomatic, political or military means. Once passage through the GIUK gap was assured, submarines and other naval units could move freely into the Atlantic from Murmansk, supported by the unrestricted operations of maritime aircraft, and the approach of NATO units could be contested long before they could come close to the sensitive areas of the Kola Peninsula and the White Sea.

Much of Soviet strategy both before and after World War II had as its aim the construction of buffer zones to keep potential invaders at a distance from the Russian homeland. Indeed, the demands made on Finland in 1939 were intended to protect the approaches to Murmansk as well as Leningrad; the subsequent Russo-Finnish war and its settlement achieved precisely those ends. On the Pacific front, strategic islands close to Japan were appropriated. There are even more obvious examples of the desire for buffer zones on land: eastern Poland in 1939, Lithuania, Estonia, Latvia, Bessarabia and northern Bukovina in 1940. In fact, the Soviet Union's partners in the Warsaw Pact can be considered as buffers between her and NATO. Thus it would be entirely consistent for Soviet strategists to wish to construct similar buffers on the maritime frontiers, and to push them outwards whenever an opportunity occurred. Beginning with the nearest to the Murmansk coast, the candidates would be Northern Norway, Svalbard (including Bear and Hope Islands), Jan Mayen, Greenland, Iceland and the Faroes.

SUMMARY

The main maritime tactical lesson learned in World War II was that the most efficient use of anti-submarine forces was in close escort and support of merchant convoys, rather than in combatting submarines in transit. However, post-war improvements to submarines, particularly nuclear propulsion and the fitting of anti-shipping cruise missiles, have outstripped advances made in anti-submarine technology. The most effective anti-submarine vehicle is now the nuclear-powered submarine, but it is better suited to defending a barrier against transiting submarines than to escorting convoys. It will still be advisable for merchant ships to travel in escorted convoys, largely for protection against aircraft and cruise missiles, but a substantial proportion of anti-submarine defence should probably be devoted to the establishment of barriers to prevent the transit of the attack submarines between their bases and the area of the convoys. An obvious place for a barrier is the GIUK gap.

The other great post-war development in submarines has been the SLBM, which gives them a new role as the most important component of the strategic deterrent. There is every reason for the strategic submarines to stay well clear of the areas in which anti-submarine forces will be operating in defence of shipping; with their very long-range missiles, they can conduct their missions without crossing the GIUK gap.

Geo-strategic features demonstrated in World War II which are still valid include:

1. the importance for the Atlantic powers of Iceland, Greenland and the Azores for the defence of the supply routes from America to Europe;

157

2. the constraint placed on the Arctic power by the GIUK gap;
3. the capability of forces based in Norway to harass naval operations out of Murmansk; and
4. the importance of the possession of Norway to a naval power wishing to operate into the Atlantic out of the Baltic exits.

Thus, for the preservation of the sea lines of communication across the North Atlantic, the key areas are likely to be the Norwegian and Greenland Seas.

BIBLIOGRAPHY

Capt. R. Beavers, 'The End of an Era', *USNIP* 833, 19–25 (July 1972).
Capt. R. A. Bowling, 'Escort of Convoy, Still the Only Way', *USNIP* 802, 46–56 (Annapolis: Dec. 1969).
Ian Campbell and Donald Macintyre, *The Kola Run* (London: Futura Publications 1975).
Paul Cohen, 'The Erosion of Surface Naval Power', *Foreign Affairs* 49, 2, 330–341, Jan. 1971.
Paul Cohen, 'New Roles for the Submarine', USNIP 835, 31–37 (September 1973.
Capt. John Creswell, *Sea Warfare 1939–1945* (Berkeley, Calif: University of California 1967).
Vice-Adm. Sir Peter Gratton, *Maritime Strategy: A Study of British Defence Problems* (London: Cassell 1965).
C. G. Jacobsen, *Soviet Strategy _ Soviet Foreign Policy* (Robert Maclehose 1972).
Robert Kuenne, *The Attack Submarine: A Study in Strategy* (New Haven, Conn.: Yale 1965).
Donald Macintyre, *The Battle of the Atlantic* (London: Batsford 1961).
Samuel E. Morrison, *History of United States Naval Operations in World War II:*
    Vol I, *The Battle of the Atlantic,* Sept 1934–May 1943 (Boston, Mass.: Little Brown, 1960),
    Vol X, The Atlantic Battle Won, May 1943–May 1945 (Little, Brown, 1959).
Major Gen. J. J. Moulton, 'The Defence of Northwest Europe and the North Sea', *USNIP* 819, 80–97 (Annapolis: May 1971).
Edward Von der Porten, *The German Navy in World War Two,* (New York: Ballantine, 1974).
Alfred Price, *Aircraft Versus Submarine,* (Annapolis, Md: Naval Institute Press, 1973).
Captain S. W. Roskill,* *The War at Sea, 1939–1945:*
    Vol I,   *The Defensive* (London: HMSO, 1954),
    Vol II, *The Period of Balance* (HMSO, 1956),
    Vol III, *The Offensive:* Part I, June 1943–May 1944 (HMSO, 1960),
             Part II, June 1944–Aug 1945 (HMSO, 1961).

* Much of the historical data in this chapter has been taken from *The War at Sea.*

Royal Air Force 1939–1945:
Denis Richards, Vol I *The Fight at Odds* (London: HMSO, 1953),
D. Richards and H.St.G. Saunders, Vol II *The Fight Avails* (HMSO, 1954),
H.St.G. Saunders, Vol III *The Fight is Won* (HMSO, 1954).

Capt. W. J. Ruhe, 'The Nuclear Submarine: Riding High', *USNIP* 864, 55–62 (Feb 1975).

Capt. R. H. Smith, 'ASW – The Crucial Naval Challenge', *USNIP* 831, 128–141 (May 1972).

Capt. G. E. Synhorst, 'Soviet Strategic Interest in the Maritime Arctic', *USNIP* 843, 88–111 (May 1973).

The Military Balance 1974–1975 (London: IISS, 1975).

Tsipis, Cahn and Feld (eds.), *The Future of the Sea-Based Deterrent* (Cambridge, Mass.: MIT Press, 1973).

Cdr. Nicholas Whitestone, *The Submarine: The Ultimate Weapon* (London: Davis-Poynter 1973).

DANIEL P. O'CONNELL

Resource Exploitation, The Law of the Sea and Security Implications

In introducing the question of international law into a discussion of the defence of the North Atlantic I think I should begin by explaining its relevance. To those unacquainted with the overall structure of international law but aware of the current struggle over the Law of the Sea, it must seem that international law is a spurious creature designed merely to make political attitudes appear more reasonable or to increase their psychological effectiveness, and that it fails in the law's traditional role of solving disputes. Although the greater part of international law functions smoothly in the traditional role, it is certainly true that in matters to do with the Law of the Sea the law suffers from the disorder wrought by political and economic changes throughout the world. The methodology of international law has been transformed and, some would say, gravely affected, for no longer can states be relied upon to act according to their legal consciences so that their practices are descriptive of the law. Rather, it it obvious that they intend deliberately to break the law in order to bring about the changes in its rules necessary for them to achieve their political objectives.

This change in the methodology of international law is of direct relevance to the present topic for two reasons. First, the law is clearly not at a discount, even if it is politically harnessed; on the contrary, it has greater significance than ever, because it is seen as an instrument of political leverage and economic change. Secondly, the shifting emphasis of its purpose equates the law not with morality but with power.

Both these aspects bear upon defence thinking. The first suggests that governments are likely to have to defend their policies by reference to law, for that is a moral and psychological requirement of our times. Gone are the days when Palmerston could threaten to use the might of the Royal Navy without even mentioning the legal basis of his claims, or even when Anthony Eden could contemplate reprisals against Italy for clandestine submarine attacks during the Spanish Civil War, without taking legal opinion on the point. The second point suggests that military power will be used to make new legal stances effective, and to prevent others from becoming effective.

International law has always been in the national armoury, even if this was not always recognized or the weapon was not always cleverly used. But today the point is obvious and no one who has had any naval operational experience in recent years doubts that the adroit manipulation of the Law of the Sea and the law of self-defence can enhance the tactical possibilities and make the defence of naval actions in the United Nations more plausible – without which tactical success could pay no political dividends. It follows that international law is a fundamental part of operational planning concepts.

This will become clearer after I have made some observations about the developments currently taking place in the Law of the Sea. The United Nations has been struggling with the political complexities of yet another revision of the Law of the Sea since 1967. The present so-called codification of the Law of the Sea, the Geneva Conventions of 1958, has only been in force for thirteen years, yet it has been broken already. There were long sessions of the Law of the Sea conference in 1974 at Caracas and 1975 at Geneva, but there seems to have been little agreement among the participants. The politics of the subject may have been responsible for the disjointed and ragged way in which the meetings were conducted, but it is none the less true that it was a deplorable diplomatic spectacle, an embodiment of the degradation of diplomacy that has occurred in recent years.

What Geneva has produced is a set of three documents, each called a Single Negotiating Text and each produced by the chairman of one of the three committees. At least we now have a text to argue about, and that is progress; up to the present we have had only a plethora of proposals, some well and some badly drafted. In the Icelandic case half the judges of the International Court of Justice dismissed these as nothing but aspirations, while the other half thought they might contain the kernel of new rules of law.

The Single Negotiating Texts are supposed to embody what the second group of judges discerned in the proposals – that is, common trends towards majority positions which could be considered to be developing rules of law. It remains to be seen whether the chairmen who drafted these texts did in fact accurately gauge the trends, and whether the recalcitrants, of whom there are still many, can be brought into line. The debate, and indeed the animosity, has by no means ended.

When we consider the defence of the North Atlantic in the middle 1970's, it obviously becomes important for us to take into account the trends in the Single Negotiating Texts to see how, if they are consolidated, they will affect the matter. Broadly speaking, international law affects maritime operations in two ways: first, it defines the areas of sea within which conduct of a particular kind is allowed; secondly, it prescribes how ships are to behave within these areas. When the rules on either of these subjects – one spatial, the other qualitative – are unclear, controversial or too generalized, possibilities exist for dispute and for clouding the issues, for plausible defence of conduct and, thus, for the promotion of operational, and ultimately, political ends.

I now propose to go systematically through the Single Negotiating Texts to see what they contain that could affect the areas within which naval operations in the North Atlantic are likely to be conducted. I have said that there are three Texts: the first concerns the machinery for international control of mineral exploration and exploitation in the seabed of that part of the high seas not subject to any form of national jurisdiction – what we shall call the 'deep seabed'; the second concerns the areas of national jurisdiction; and the third deals with the conservation and preservation of the marine environment, or anti-pollution jurisdiction.

The subject matters of all three are obviously interrelated, and a great deal more co-ordination will be necessary before they can be integrated in a comprehensive scheme. For example, pollution control appears in all three Texts, but in incompatible ways; in some cases the extent of the coastal state as pollution authority is in inverse ratio to the extent of its authority in the various zones in other matters. I shall not say much about pollution because this is the one matter on which I expect a breakdown of the present Texts under the pressure of maritime-power interests, but I will summarize the points briefly.

There is a widespread assumption on the part of the non-maritime states that wanton pollution of the seas occurs, due to the unwillingness of the shipping nations – in the interest of furthering capitalist aims – to police their own vessels. The remedy they see is the seizure of authority over ships passing their coasts at distances within which pollution, due to currents or other factors, could affect their interests. In other words, they see a legal remedy for a social evil. But the truth is that it is not the law that is at fault here. The Inter-Governmental Marine Consultative Organization conventions on oil pollution and the Dumping of Waste at Sea Convention are quite adequate to deal with the problem, even if they leave it to the flag states of the ships to regulate the matter. If these conventions are not working satisfactorily this is due to two things: first, the inertia of governments, which makes accession to them inordinately slow – and here the fault is not with the shipping countries, but with the coastal states, who have not yet passed laws to give effect to these conventions, mainly because of the great difficulties involved. Secondly, there is the problem of evidence of fault in the master of the ship, which is not going to be solved by coastal states asserting their authority over passing ships, since their judicial systems are often no better equipped than those of the shipping countries to gain jurisdiction over the foreign ship or to try the issues, and their maritime forces are inadequate either to arrest the ships or to collect the necessary evidence.

There are exceptional geographical areas where the general considerations may be different from those I have just stated. Canada, for example, considers the Arctic to be such an exception and asserts coastal state authority there, but Canada's own arguments are based on the geophysical and ecological peculiarities of the area and do not bear upon the question of pollution of the open oceans generally.

I turn now to a consideration of the overall scheme of the Single Negotiating Texts. Basically, a distinction is drawn between the area of national jurisdiction and the areas beyond national jurisdiction. Within this second area an international authority is to be made responsible for regulating exploration and exploitation. Within the area of national jurisdiction the coastal state is exclusively competent, and this area will extend, if the text is accepted, to 200 miles, or to the median line when there is less sea-room than 400 miles. Of these 200 miles, the first 12 will be territorial waters, within which the coastal state is sovereign; the remaining 188 will be an exclusive economic zone, within which the coastal state will have the sovereign right to explore, exploit, conserve and manage the renewable and non-renewable natural resources of the seabed and waters, and will have (I quote in order to underline what I said earlier) 'jurisdiction with regard to the preservation of the marine environment, including pollution control and abatement.'

When one looks at this scheme from the defence point of view it has its positive and its negative aspects. On the positive side it aims to eliminate the ambiguity at present associated with the existing 200-mile claims – which are, if not actually territorial claims, in some cases tantamount to them – although its success in this aim is not a foregone conclusion. If it does succeed, and the 200-mile claimants all accept the proposed rule, naval forces, with their associated air activities, will have the advantage of the same freedom of action and manoeuvre beyond 12 miles from the coast as they have had hitherto; submarines could transit anywhere outside the 12-mile limit submerged, whereas within the 12-mile limit they may transit only in the exercise of the right of innocent passage – that is, on the surface and showing their flag. The negative side of the scheme is the vagueness of the Text about the extent of the coastal state's powers, which might be extended to control even the type of propulsion of ships in the exclusive economic zone.

The fear of the shipping powers is that, once coastal states become entitled to assert their authority over shipping for reasons of conservation and protection of the environment, there will follow that phenomenon, familiar to international lawyers, called 'creeping jurisdiction'. Within the 200-mile zone is it proposed that coastal states will have jurisdiction over natural resources, but this will entail authority over ships which threaten the degradation or destruction of these resources. So pollution control follows as a matter of course from exclusive economic rights. But will it not then also follow that the coastal state may forbid the presence of ships which could cause inordinate risk, such as ships not constructed in a certain way? Will control stop short of the passage of nuclear ships? And, if these are nuclear warships or nuclear submarines, will there be a conflict of interest between coastal-state preoccupation with resources and great-power preoccupation with strategy? Will it stop even there? For example, it might be argued by a coastal state that fall of shot, missile-splash or the dropping of Jezebel sonar buoys could be regulated, under the power to conserve the environment within the exclusive economic zone.

The really substantial bone of contention at the Law of the Sea conference was the question of transit rights through straits. Upon this question the NATO and Warsaw Pact powers found that their positions coincided, although their strategies differed. The United States, for example, begins her argument in favour of freedom of passage with the contention that no territorial sea claim beyond 3 miles is valid; in all straits that are wider than 6 miles there should be a belt of high seas subject to high-seas passage. If it is agreed to the satisfaction of the United States that waters within straits, that were fomerly high seas, will remain subject to transit rights equivalent to high-seas passage, then the United States will accept a change in the law of territorial waters to 12 miles. The Russians, who have had a 12-mile limit, for many purposes, since 1910, start off with the more correct argument, in my opinion, that there is a special regime of transit rights through straits, different from and greater than that of innocent passage through the territorial sea, and that it is only a question of rewording this, not a question of changing the law. Tactically this has put the Russians in a better negotiating position, although they have not made notably more or less headway than the United States.

It is unnecessary to go further into this question of passage through straits except to draw attention to any possible application it may have to the Denmark Strait. Because of the territorial sea limit of 12 miles this would not fall within the regime as set out in the Single

Negotiating Text because the right of passage in the exclusive economic zone, which is what the Denmark Strait would become, would remain high-seas passage, subject to the doubts I have mentioned on
environmental grounds.

So far, then, it would seem that the present draft has not greatly
altered the situation in the North Atlantic inasmuch as it affects the
deployment of surface ships, submarines and maritime reconnaissance
aircraft. But this is not the end of the matter, because the defence of
the North Atlantic is also a matter of devices and perhaps of installations. One must look into the future, and it is here that the Single
Negotiating Texts cause one to ponder.

Let us begin with the general question of scientific research because
this would seem to cover the oceanographic activity which naval
powers engage upon with a view, for example, to establishing data for
sonar systems. Within the territorial sea, that is within 12 miles, it is
to be formally proclaimed that no act may occur that is aimed at col-
lecting information to the prejudice of the defence or security of the
coastal state. But it could be a matter of which side a passing warship
belongs to, whether the naval scientific research it conducts offends
this prohibition or not. It would seem that streaming devices in
the territorial sea would be outlawed under the general propositions
about continuous and expeditious transit, and it is in fact stated that
passage is prejudicial to the peace, good order or security of the coastal
state if it is accompanied by research or survey activities of any kind.
One even wonders if the use of variable depth sonar by ASW escorts
would fall foul of this rule.

Now turn to the exclusive economic zone. Here it is blandly stated
that the consent of the coastal state shall be obtained in respect to any
research concerning the exclusive economic zone undertaken there.
The draft goes on to say that the coastal state shall not normally with-
hold its consent in the case of a scientific institution conducting purely
scientific research, but stipulates that the results of the research must
be shared with the coastal state. It is obviously questionable whether
naval research concerns the exclusive economic zone, even if it is
directed to water temperatures rather than to fish stocks, because the
two are interrelated.

So much for scientific research in the areas of national jurisdiction –
which, remember, would cover all the access routes from the Atlantic
to the Arctic in a mosaic of interlocking 200-mile zones. What would
the situation be in the area beyond national jurisdiction, such as the
wide spaces between Ireland, Iceland and Newfoundland? Well here
we turn to the first Single Negotiating Text, which is concerned with
international control of deep-seabed mining. This area is proclaimed

to be the common heritage of mankind. Activities (i. e. mineral exploration and exploitation) are to be subject to regulation and supervision by an authority called the International Seabed Authority, which is structured like the United Nations, with an Assembly, a Council, a Technical Commission, an Economic Planning Commission, an Enterprise and a Court, each with its voting patterns which will inevitably produce political alliances, as has happened in the United Nations.

No mineral exploration or exploitation is to be permitted except under licence from the Authority. How states or their public enterprises or private companies will actually block exploration remains to be seen, since everything depends on votes and the Enterprise itself can undertake activities as a sort of super-nationalized industry. But the details do not matter for the present; all we are concerned with is the possibility that the seabed beyond the mosaic of 200-mile limits will be divided up into blocks wherein someone or other will have rights, and that the Authority will have an undetermined power of regulation. The relevant question is: how will this affect defence activities?

In the North Atlantic the answer is that they will probably not be directly affected at all. The problem, however, is the political use that could be made of the ill-defined situation by a state concerned with reducing another's defence capability in the area. For example, the Text prescribes that stationary and mobile installations are to be erected, emplaced and removed under the rules of the Authority, and are to be used exclusively for peaceful purposes. This is clearly not directed at military devices on the seabed, but could be made to appear so for political reasons. Scientific research is to be carried out exclusively for peaceful purposes and for the benefit of mankind as a whole. Again, this is not intended to catch military research on the seabed unconnected with exploration for minerals, but could be made to do so.

I do not want to make too much of the defence implications in the first Single Negotiating Text, especially in the North Atlantic context, because I am sure that the naval powers think that they could live with them, but it would be wrong not to recognize that the designation of the high-seas area as the common heritage of mankind contains the germ of neutrality for the oceans; whether the powers inhibit its progress or not is another matter.

I now turn to the implications of these developments for the naval use of surveillance systems and underwater weapons systems. From a practical point of view we would here be concerned with three things: the laying of acoustic arrays on the seabed, the location of deep minefields or some system of anti-submarine torpedo-launchers, and towed sonar arrays. In present international law the freedom of the seas means that no fundamental political obstacle exists, at least to the use of arrays, although the location of minefields does raise questions about notification under the Hague conventions on the Laws of War. But the grip which an International Seabed Authority might progressively gain could raise questions, and merely the fear that they might be raised could cause caution among the Great Powers regarding decisions which the Authority might take even in quite innocent matters. For one thing, whether arrays benefit from the freedom to lay cables might be questioned.

This type of question has already arisen in connection with the continental shelf, and has influenced the attitudes of the powers in both the Law of the Sea Conference and the Disarmament Conference. The question is whether the exclusive sovereign rights of the coastal state with respect to its continental shelf entitles it to the exclusive right to lay acoustic arrays and implant weapon systems thereon. The Soviet Union, after a period of uncertainty, appears to have come down in favour of the continental shelf being as free as the high seas for this purpose, so that any state may use the seabed beyond the territorial sea for military purposes, other than the emplacement of weapons for mass destruction, which is outlawed by the treaty on that subject. But other naval powers may take the opposite view and argue that the continental shelf is national territory. Obviously the legal difficulties involved in continental shelf boundary negotiations influence the matter.

The continental shelf is likely to be superseded by the exclusive economic zone, at least so far as the North Atlantic is concerned, so that this controversial question could become a more general one, and perhaps a more difficult one considering the extent of coastal-state control that would be exercised over the waters as well as over the seabed. The use of military devices would then benefit less from the freedom of the seas.

The same could occur in the case of towed arrays. These are not only useful in extending the range of contact for purposes of anti-submarine warfare but are also likely to be used to locate fixed arrays. An electronic battle is a possibility in which the competition could take on the aspect of scientific research, which would bring it within the scope of the draft articles.

It may be thought that the rules of international law are likely to have little influence upon the conduct of naval operations in an area of vital strategic significance and would in practice be ignored. It would be wrong, however, to be too sceptical on this score, because we are not concerned only with what would happen in the event of war but what happens in the shadow of the deterrent and what would happen in a period of tension. The adroit manipulation of legal issues, bearing

167

as they do upon the political attitudes of the countries surrounding the North Atlantic, could lead to one side or the other gaining an advantage in the process of escalation. This is why naval staffs take what is occurring at the Law of the Sea conferences so seriously.

So far I have spoken of geographical areas of possible conflict, but I should like to conclude with a few observations on the role of international law in the situation of tension. Here the matter is regulated by the United Nations Charter, which outlaws resort to force but at the same time reserves the inherent right of individual or collective self-defence against an armed attack. All escalation policy takes into account the need to adopt a stance of self-defence which is plausible. The problem here is to know how serious the hostile intent is in a threatening force, and to gauge the moment when it is to be translated into a hostile act. Much thought is being concentrated upon the problem of pre-emption, and attempts have been made to refine identification of the point at which hostile intent becomes hostile act. These efforts have not been very successful, but we are left with one fundamental important difference between Soviet and NATO thinking: the Soviet Union, as Admiral Gorshkov's articles illustrate, believes in a saturation and high-level use of force; she appears to have no place for limited war, periods of tension or graduated force. The NATO navies have certain more or less refined answers to this attitude, although some might say that excessive caution would emasculate them in a showdown.

The implications of this debate are not of immediate concern, except to say that the advantage here may lie with NATO in the North Atlantic. This supposes that the political authorities in the Soviet Union would not in fact authorize the Soviet Navy to use its single-strike capability unless they were prepared for total war, and that navy has no doctrine for use of anything of a lower order. NATO forces may prove to be better exercised in the politics of escalation and hence in the tactics of marking and countermarking. The law plays a role here, perhaps not a major role but none the less an essential one, in regulating the procedures of harrassment and governing the areas within which operations can be mounted to political advantage.

The significance of the introduction of international law into defence-thinking and policy-making cannot be either over- or under-emphasized; it is one of the elements in the compound. Confrontations over issues such as fishery rights can threaten defence relationships, so that agreement upon the law can become a matter of some priority. The important point for defence planners to recognize is that the legal factors in the strategic equation are constant and cannot be ignored.

JOHAN JØRGEN HOLST

The Strategic Importance of the North Atlantic: Some Questions for the Future

THE CENTRALITY OF THE CENTRAL BALANCE

The strategic importance of the North Atlantic is to a significant degree a function of the future of the central balance between the Soviet Union and the United States. That is in turn significantly influenced by technological developments. The improvements in the accuracy with which intercontinental missiles may be expected to hit fixed targets may render the fixed land-based ICBM obsolete. In any event, it seems reasonable to expect that a relatively larger portion of the strategic missile eggs will be placed in the submarine basket. What is not so often remembered is that force deployments at sea inevitably carry with them geo-political consequences. It is in the area of such consequences that the greatest uncertainties concerning the future strategic importance of the North Atlantic lie.

The major part (at present around 70 per cent) of the Soviet strategic missile-carrying submarines, SSBN, will continue to be based with the Northern Fleet on the Kola peninsula. The increased range of the SS–N–8 missiles, which are deployed in the DELTA-Class submarines, will permit rearward patrols in the Greenland, Norwegian and Barents Seas, though it is too early to extrapolate future deployments from the present low-level deployment of a couple of DELTA-class submarines to the areas in question. Moscow may prefer to achieve greatest deterrence by avoiding a concentration of her land-based and submarine-based missile power in the same Arctic threat corridors, and attempt instead to threaten the United States from different directions and by systems of different ranges. The latter consideration may affect Soviet incentives for not retrofitting YANKEE-class submarines with SS–N–8 as replacements for the shorter range SS–N–6. The development of two successor missile systems indicate Soviet interests in a differentiated SLBM force. Moreover, Moscow would have an interest in not concentrating her force in a rather limited ocean space: the shallow Barents Sea might not seem ideal for SSBN patrols.

A great deal would seem to depend on whether Moscow will choose

169

to change the present operational pattern of keeping a relatively modest number of SSBN's on station at any time – some 16 per cent of the DELTA-class submarines and 8 per cent of the YANKEE-class submarines in the Northern Fleet. Little is known about why the Russians keep such a small portion of their SSBN force on station in peacetime. Problems of command and control may suggest part of the answer, but presumably such constraints may be overcome. The lack of overseas bases and tenders offer only a limited explanation. Technical problems connected with maintenance and overhaul are probably among the relevant factors. Moscow may look at the SSBN force as a withholdable force to be used for second-strike and wartime bargaining purposes, and may also expect a strategic warning, permitting surge deployments in a crisis. From such a perspective the increased range of the SS–N–8 would enable Moscow to deploy rapidly.

What would be the geo-political consequences of a rearward Soviet SSBN deployment? We should expect Soviet SSN patrols as well as surface units to protect the SSBN deployment areas by forward patrols in the Norwegian Sea. The Soviet negotiating posture on demarcation of the continental shelf in the Barents Sea and attitudes to the exploitation of offshore resources in the Barents Sea by Norway and her Western allies will probably be influenced in large measure by naval interests in maintaining the SSBN deterrent. The explicit or tacit negotiation of submarine sanctuaries could become a priority objective which would have obvious implications for the political and economic sovereignty to be exercised by Norway over the resources in the area.

The increased range of American SLBM's will decrease United States' dependence on being able to maintain SSBN patrols in the Norwegian Sea, an area in which the Soviet Union can mount an impressive hunter-killer threat to SSBN patrols. How will this development affect American naval interest in the North East Atlantic in general? Presumably that interest will be associated primarily with general-purpose support to Northern and Central Europe. The Soviet power to intercept the North Atlantic supply routes constitutes a serious challenge to the umbilical cord of the Atlantic Alliance. A Soviet dominance of the Norwegian Sea would have very serious consequences for the perceived credibility of the American guarantee to Western Europe.

The probable fact that American and Soviet SSBN patrol areas will not be superimposed on each other in the North East Atlantic in the years ahead may be an important stabilizing development from the point of view of the central balance, although British and French SSBN patrols will continue to make the situation somewhat ambiguous. Norwegian security stands to gain, however, from the existence of European strategic interests in the Norwegian Sea. The first *Trident*

SSBN will be operational in 1979, and American ASW efforts will presumably still make up a forward element of United States continental defence.

THE REGIONAL BALANCE AND THE IMPORTANCE OF THE CENTRAL FRONT

Continued American presence in Iceland and the ability to reinforce Keflavik and air-bases in Southern Norway rapidly with long-range tactical fighters permits the Western Alliance to set up a region of air dominance over large areas of the Norwegian Sea and thus deny the Soviet Union naval dominance by surge deployments. The ability to commit carrier groups, hunter-killer submarine forces and maritime patrol aircraft to the North Atlantic in an emergency will continue to form a redressing element in a situation which, on the face of it, would seem to fall within the sphere of Soviet dominance. The capabilities will in principle be available, though the constraints and risk assessments which will apply in specific circumstances cannot be predicted precisely. That same caveat, of course, would apply to the calculations of a would-be aggressor.

The numerical reductions in American naval ships will inevitably affect the long-term naval balance. However, in spite of the reductions in combat units the United States Navy has been able to maintain a claim on a substantial portion of the total defence budget. The modernization programme now under way has produced some important qualitative improvements which have to be weighed against the reduction in ship platforms.

The ability to protect the sea lines of communication (SLOC) is to some degree a function of the number of naval vessels available for the task. There has been a significant reduction of the United States Navy to below 400 units. However, at the same time the average size of tankers and cargo carriers has increased significantly, thus reducing the number of ships needed to carry a certain load. There will be fewer units needing escort protection, but fewer vessels will need to be sunk to achieve significant attrition. There may consequently be a case for using a larger number of smaller transport vessels in future transatlantic convoys. In this connection it is necessary, however, to consider the contingencies which might create a Soviet need to control the Atlantic SLOC. Such needs would presumably be most likely to occur in a long war or in the context of a 'phoney war' of build-up in Europe and attrition at sea. In the former instance the port facilities at the receiving end in Western Europe might well be the most vulnerable nodes in the total reinforcement system.

In the event of a major reduction in American military presence in Central Europe much of the burden of Western defence would be transferred to the transatlantic SLOC. In this context, it should be pointed out also that the security of Northern Europe is intimately linked to the ability of NATO to cope with escalation in the centre. Should the latter deteriorate, Northern Europe could no longer expect the umbrella of deterrence to extend to the peripheral areas with sufficient credibility in the calculations of potential aggressors.

The naval forces of the states of Western Europe would contribute significantly to the ability of the Western Alliance to maintain and protect the SLOC in the Atlantic and in regional theatres. How likely are scenarios in which that capability really matters? Such questions do not permit precise answers, but the ability to protect SLOC to Northern and Central Europe may figure as an important option in crisis management and bargaining. The absence of an ability to maintain the SLOC could presumably be exploited by the dominant land power to influence expectations about the outcome of regional conflicts.

SOVIET NAVAL POWER AND ITS POLICY IMPERATIVES

There is sometimes a propensity among Western observers to dramatize Soviet naval operations in the North Atlantic. What has taken place over the last decade is not so much a dramatic build-up as a move to forward deployment and operations. The number of units has even decreased, but the quality has improved considerably, both in terms of endurance and firepower. Many questions remain on the regional and global impacts of the Soviet move to the sea. To what extent will the size and configuration of the Soviet Navy create policy imperatives which will drive Soviet foreign policy in specific directions? Navies have traditionally carried with them geo-political imperatives and interests. The world-wide spread of Soviet naval power may influence the priorities and interests in Soviet world policy. Such questions may be of particular importance and anxiety in the littoral states of the North East Atlantic. How will weaknesses and vulnerabilities in the Soviet naval posture in the Norwegian Sea influence Soviet decision-making in a crisis? What territorial imperatives will operate in a crisis? What existing instabilities will affect the risk calculations of the relevant actors in a North European crisis? How will naval interests affect Soviet decision-making and bargaining behaviour regarding management regimes for ocean resources in the North-East Atlantic? How will the heavy Soviet investments in distant-water fishing affect Soviet policy towards those LDC's which will be unable to exploit the maximum sustainable yield within their own resource zones? How will competition for access affect Soviet-Japanese relations, and by derivation Sino-Soviet relations? Will the Soviet Navy become an instrument deployed in support of Soviet interests in access to ocean resources and waterways? To what extent will American and Soviet interests coincide, and to what extent will they come into conflict in the emergence of a new law of the sea? How will general consensus affect specific behaviour in particular areas, such as the North Atlantic? Such questions cannot be answered in any definitive fashion: they are largely conjectural and depend upon future events. But they indicate some of the paths of potential conflict and co-operation in the North Atlantic in the years ahead.

SOME FUTURE UNCERTAINTIES

If we look even further ahead, uncertainties grow broader and more numerous. Will the Arctic develop into another Mediterranean; a lifeline of communication among the industrialized countries of the northern hemisphere? To what extent will an opening up of the north generate a kind of raw materials autarchy in the industrial countries, and how will access to Arctic on-shore minerals and off-shore hydrocarbons affect the distribution of bargaining power within a new economic order? Will it promote or impede the development of global equity? Will it establish the Soviet Union as a major factor in the international economic order? What stakes and interests will prevail in Moscow, Ottawa, Washington, Copenhagen and Oslo? Will the opening up of the Arctic provide opportunities for co-operation across the political divisions of the East-West conflict? To what extent should we expect novel patterns of conflict to emerge and how will conflicts relating to European politics and world power relations be affected? Will the ecological vulnerability of the Arctic compel the littoral states to co-operate so as to avoid disaster? Will man be able to comprehend the systemic effects of marginal incessions and modifications? How will the cyclical forces of nature affect conditions for human activity in the high north in the years ahead? What technological incentives can be derived from and discerned in the present situation?

In the short term security planners are trying to comprehend the messages contained in the evolving strategic situation in the North-East Atlantic. The Soviet power of interference and interception seems bound to affect risk calculations in relevant capitals in times of crisis. Will reinforcements to Norway be forthcoming when high risks of escalation must be taken into account? Will the Soviet Navy be able to establish the same kind of escalation control in the Norwegian Sea during a crisis in the late seventies as the United States Navy managed to

173

provide during the Caribbean crisis in 1962? What time factors will apply in containing the submarine threat to transatlantic reinforcements to Northern Europe? What measures can be implemented for purposes of designing around the Soviet submarine fleet? Does positioning heavy equipment beforehand provide a partial answer? Should greater emphasis be put on a rapid-confrontation option designed to increase the shared risk of escalation by bringing allied hostages to the front lines at an early stage? Would a rapid transfer of long-range fighters and ASW aircraft to bases in Southern Norway provide a rapid deterrent and counter to Soviet naval attempts at sealing off the GIUK gap?

STRAINS ON THE EQUILIBRIUM OF POWER

The question is sometimes raised whether the Norwegian and Danish reservations about foreign bases in peacetime are still functional policy constraints, in view of the forward operations of the Soviet Navy. It should be remembered, however, that these self-imposed restraints, which have been observed since 1949, have become integral elements of the regional equilibrium of Northern Europe and, by extension, of the security order obtaining in Europe regarding the East-West balance. Any change in the position of Norway and Denmark would alter the existing *status quo*. It seems unlikely also that the United States should be willing to position troops in a forward area overseas while American commitments are being reassessed and the attempt to arrive at rules of superpower conduct predicated on the exercise of reciprocal restraint. The adjustments will, it seems, have to be found and implemented within the confines of the present balance of commitments and restraints operating in Northern Europe. But will Soviet naval power be employed for purposes of exercising pressure on the littoral states to move away from the current premise of an established equilibrium? We should distinguish here between power focused on particular circumstances and general background 'reminders.' What would be the appropriate NATO response to such tactics? Could they be distinguished with certainty? What tactical gains would be associated with particular ambiguities? Would the deployment of a United States carrier task force to the North-East Atlantic, and based in the United Kingdom, Holland or Northern Germany, constitute an appropriate deterrent or response to such exploitations of the weight of the Soviet Navy? Would such a move exacerbate competition rather than allay fears? Perhaps such moves should be maintained rather as options to counter large and sustained Soviet naval deployments to the North Atlantic which aim to demonstrate an ability to deny sea control to the NATO forces.

ANTI-SUBMARINE WARFARE

The primary threat to Western sea control will for a number of years be the impressive Soviet submarine force which could be deployed to the North Atlantic. However, even nuclear-powered submarines have their problems. If they move at any significant speed the propulsion force will increase the loss of energy in the form of acoustic noise and make them vulnerable to passive detection. Under favourable conditions such detection may be possible at impressive ranges. Soviet submarines are notoriously noisier than their Western counterparts. If the submarine is operating in water illuminated by active sonar, it will probably have to move at creeping speed in order to avoid detection. However, temperature and salinity gradients – and the unpredictable variations in these gradients which are frequently encountered in the coastal areas above the continental shelf – will cause the effective sonar range to vary considerably at different times and places. For a long time to come ASW seems unlikely to reach the level of expected effectiveness now envisaged for weapon systems operating in the air or on land.

The weapon-carrying capacity of submarines is limited, as is the number of missiles which can be fired and guided in succession from submarine platforms. If fired at long ranges the missile systems will have a limited ability to discriminate between escort and transport vessels in a convoy. Because the missiles will have to rely on some kind of homing device they will in principle be vulnerable to countermeasures in the form of misleading radiation from the targets or as false targets or decoys. The use of long-range weapons would create increased demands on the command and control system. It would also tend to blur established distinctions between tactical and strategic missions and requirements. The primary task of nuclear-powered hunter-killer submarines will be under conditions of present and future technology, to go out hunting for ballistic missile-carrying submarines rather than to go after shipping and surface combatants.

Development of 'smart' mine technology may provide effective options for the establishment of barriers across strategic choke points, such as the GIUK gap, and hence bring a new uncertainty to any race for closure and entry.

A NEW LAW OF THE SEA

The emerging Law of the Sea appears to confer extended sovereignty on the coastal states in the exploitation and management of offshore resources, particularly fish, oil and gas. The extensions of the old territories are great, as exemplified, for example, by Norway's claims to an

175

area some five times larger than herself. A noticeable discrepency will emerge between claimed rights and policing structures available to the claimant states. How will the system of allied commitment to the defence of the old territory survive the impact of intra-alliance conflicts about access to offshore resources? Will excessive claims breed attitudes and expectations which drive governments into adopting short-term positions which will produce results inimical to their long-term security interests? In the absence of a negotiated general treaty emerging from the third United Nations Conference on the Law of the Sea, UNCLOS III, should we expect a competitive scramble which will bring traditional allies, such as Norway and Denmark, into conflict with Britain and the Federal Republic of Germany? What mechanisms will be available to contain the dangers of a generalized Cod War?

Will the littoral states on the North Atlantic prove able to arrange unilateral actions of extension so as to minimize the scramble and greatly increase the chances of avoiding intra-alliance confrontations at sea? But how would the Soviet Union and her allies, Poland and East Germany, react to exclusion from traditional fishing grounds? Would the establishment of 200-mile exclusive economic zones (EEZ), in fact ensure better management of scarce protein resources? The prospects are not that promising in view of the fact that some of the important breeding grounds for significant stocks, such as the Arctic cod, are outside the prospective EEZ. Would the fishing effort be diverted to and concentrated in such vulnerable areas in consideration of long-term supplies? Would unilateral extensions create an environment conducive to responsible regional management at the level of the regional fisheries commissions for the North-West and North-East Atlantic for purposes of preserving the stocks? What concessions on historical rights and preferential access have to be made in order to ensure the establishment of complimentary national and regional management regimes?

The imperatives of resource politics and security policy may come into sharp conflict in the future, particularly when processed through ideological domestic politics. The third Cod War off Iceland is the most recent example of such dynamics and the intransigence that they breed. The strategic importance of Iceland to the Alliance cannot, of course, be measured in cod, kronur or pounds. However, national governments seem incapable of co-operating on functional areas to arrive at rational solutions. Iceland had few economic options in the short term and consequently very little room for manoeuvre; Britain seemed restricted mainly by considerations of prestige and marginal electoral politics. Had the pro-NATO government in Reykjavik broken down, the domestic political process might have brought the Alliance membership issue to the fore on an emotional wave of anti-British (and NATO) sentiment.

The United States is the protector of Icelandic security. Consequently, American inactivity in the presence of what was perceived as a major challenge to Icelandic economic security by Britain may lead to public questioning of the credibility of the American guarantee, especially as it is generally believed that American pressure caused Britain to back down in the second Cod War.

Should the American Icelandic Defence Force (IDF) be compelled to leave, the vacuum thus created could cause pre-emptive fears to dominate crises arising in the North Atlantic – fears of losing a race to get to Iceland first. The position of Norway could become much more tenuous, and even the security of Britain would deteriorate. The ability of NATO to maintain surveillance of the surface and submarine traffic in the North-East Atlantic would suffer a serious blow. Should the Russians win a race to Iceland, the whole North Atlantic sea route between North America and Western Europe would be in jeopardy.

Norway faces novel problems of co-existence with the Soviet Union in strategically sensitive areas in the North. The negotiations about the demarcation of the continental shelf in the Barents Sea constitute the most immediate test case. Competing interests are dressed up in competing legal arguments. But the bargaining is a political process of unequal actors. How can Norway bring countervailing power to bear on the outcome? How can she do so without making concessions on access to offshore resources which would increase Soviet fears of great-power presence in the immediate vicinity of key base areas and across strategic transit routes? The issues will be compounded by the legal tangle which may be created over the position of the continental shelf surrounding Svalbard. Here we have only witnessed the initial skirmishes with the inevitable accompanying difficulties of differentiating tactics from substance, insurances from interests and options from positions. The future is in the making, but its gestation will prove both complex and challenging.

JOHAN JØRGEN HOLST

The Strategic Importance of the North Atlantic: Norway's Role and Options

The preceding chapters have outlined the principal perspectives and activities which will determine the pattern of cooperation and conflict in the north-east Atlantic in the years ahead. The actual structure and course of events are, of course, not predictable in any absolute sense. Prognostication must be contingent, and the imponderables are many. However, it seems reasonably clear that Norway's position will be a critical one to the future shape of interstate relations in the European north. Hence, the present chapter will attempt to outline some of the major policy issues on the Norwegian agenda with respect to the north. The purpose is not to prescribe particular policy decisions but rather to establish a framework within which such decisions should be evaluated, focusing on the key connections and ramifications. The approach is deliberately normative.

POLICY OBJECTIVES

As a point of departure it seems necessary to point out that the great power interests that converge and cross in the north-east Atlantic demand a much wider perspective than what may be suggested by the pursuit of Norwegian goals of possession. It is important to define the national interest also in terms of the milieu goals which seem desirable from the point of view of international security.

It may be useful to list a set of Norwegian policy objectives with respect to the developments in the north-east Atlantic (including the Barents Sea area):

1. Maintenance of a stable and viable balance of power.
2. Preservation of a state of low tension.
3. Establishment of a viable longterm management scheme for ocean resources.
4. Protection of the preferential rights of the coastal population to the protein resources of the coastal waters.
5. Protection of the fragile ecological system of the far north.

6. Construction of an equitable and acceptable legal regime for the ocean areas involved.
7. Exercise of credible and equitable enforcement of Norwegian rights and obligations.

In the pursuit of these objectives specific care must presumably be taken to avoid constellations which would by implication damage the prospects of attaining the objectives in the first instance. The objectives may at times pose conflicting requirements, and when aligning policy in such circumstances it will be necessary to maintain an overall perspective. The dangers of sub-optimum outcomes are always present, particularly in view of the emotive potential of some of the possession goals, and the partisan competition for the support of marginal electoral groups in a finely balanced parliamentary situation. The need for intellectually strong and courageous foreign policy leadership will grow as the milieu goals which lack electoral constituencies may conflict with the possession goals of domestic groups with considerable electoral leverage. In small countries confronted with prospects of considerable acquisitions there is inevitably a danger of an 'arrogance of weakness' based on legal formalism and self-indulgent moralism. In the final analysis the issues involved have to do with allocation, with who gets what, when, and how; i.e. they are political issues which require political solutions. Such solutions have to be sought in the context of an overall framework.

Norway borders on the Soviet Union and is allied to the United States. Such is the simple but inescapable premise for Norway's integration into the East–West equilibrium. The NATO connection provides countervailing power and influence to that of the superpower neighbour. Norway has attempted to preserve confidence in her own independence by contributing to an arrangement whereunder the regionally dominant power remains contained within a larger structure of constraints. Hence, it will probably remain a Norwegian objective to avoid exclusive Soviet–Norwegian bilateralism in the north, particularly such bilateralism which would contain the seeds of conflict and confrontation with Norway's primary allies in the security field.

By the same token Norway has observed the particular restraints which small powers have to take into account so as not to provoke their great power neighbours. Such adjustments are made not out of subservience to the will of others, but rather out of enlightened self-interest in avoiding both cause and pretext for external pressure. There is a fine line to be observed here between prudence and appeasement. Norway has attempted to follow a policy of prudence by balancing measures of deterrence and reassurance in her security policy. She will confront the need to extend a credible prudential policy to her management of the ocean areas in the north.

The primary reason for the difficulties is, of course, the spatial juxtaposition of strategic interests and resource interests in the Barents Sea area. The Soviet Union will no doubt act to preserve the egress of her Northern Fleet and the access of her SSBN force to its patrol areas. Norway is a major maritime nation and shares the Soviet interest in preventing a creeping territorialization of ocean space and in maintaining a high seas regime in the EEZ and beyond. Norway may confront the need, however, to resist expansive attempts at insurance on the part of the Soviet Union, attempts which carry the centrifugal suggestion of a Soviet dominated condominium. By the same token Norway's ability to resist such centrifugal suggestions will constitute an important precondition for a continued low profile presence of the major powers of the West. It is not, of course, a question of Norway's becoming a neutral buffer state between East and West in the north. But Norway can presumably perform certain stabilizing functions on behalf of the international community by offering an alternative to direct competition between East and West in a strategically sensitive area. In many ways the task constitutes an extension of the policy of prudence pursued by Norway in her security policy since joining NATO in 1949. That policy presupposes Norwegian participation in the Western security system.

CRITICAL POLICY AREAS

We can identify five policy areas which are likely to dominate the Norweigan foreign and security policy agenda in the short term. The short term is critical because it will establish the basic contours of the long-term framework.

1. Designing around the interdictive potential of the Soviet Navy in the north-east Atlantic in order to preserve options for reinforcement in an emergency.
2. Negotiating fishing agreements with the states most affected by the establishment of a Norwegian 200 mile EEZ.
3. Negotiating a quota system for the major species in the fishing areas of the north-east Atlantic.
4. Negotiating a division of the continental shelf in the Barents Sea between Norway and the Soviet Union.
5. Obtaining international acceptance for a responsible resource management regime in the areas around Svalbard.

The present discussion is limited to an examination of the major considerations involved. It will not deal with the details of implementation. It is recognised, of course, that the process of implementation is likely to require some modification and adjustment of the principles.

Reinforcements

Norway is dependent on rapid reinforcement by national and allied forces in an emergency. That requirement is also defined to an important extent by the policy decision of not permitting the stationing of foreign troops in the country in peace time. It has sometimes been suggested that what is needed above all is to get a few American hostages to the front line quickly so as to confront the attacker with the need to shoot at Americans in order to conquer Norwegian real estate. However, if the hostages are not part of a real fighting force, their inability to affect the outcome on the ground will transfer the onus of escalation to Washington. The issue is reminiscent of the discussion about the demonstrative first use of nuclear weapons to evince will. The difficulty is the same; the demonstration may very well appear as a desperate gesture demonstrating not so much resolve as the absence of real effective alternatives for resistance. If the opponent chooses to ignore it, the defender still has to decide on what to do next, i.e. look around for effective alternatives.

In the Norwegian context what is important is to preserve a credible capability for resistance until such time as outside reinforcements can be brought in. Several measures suggest themselves. They include an improved Norwegian ability to bring mobilized brigades from Southern Norway to the north in an emergency. For that purpose the equipment of some of the mobilization brigades may be stored in North Norway. Such a program is under way. The transfer of STOL (short take-off and landing) transport aircraft from allied sources could significantly improve the strategic mobility of the Norwegian Army in an emergency. Similarly, the transfer of air defence fighters could materially extend the holding time of Norwegian forces in the north. Heavy equipment for allied forces may be pre-stocked in Norway. The Norwegian anti-invasion defences may be improved by equipping the new fighter, F-16, with anti-shipping missiles (e.g. HARPOON, PENGUIN, MAVERICK, etc.), guided projectiles for the coastal artillery, improved mines, etc. Similarly, improved airfield protection by active (ROLAND II) and passive measures will enhance the reception capability. The transfer of long range fighter bombers to Norwegian and Icelandic airfields is likely to reduce the Soviet capability to intercept transatlantic ocean-going reinforcements to Norway.

Access to Fishing Grounds

Norway will acquire exclusive rights to the resources in some of the richest fishing grounds in the world by the establishment of the 200 mile EEZ.[1] Historical fishing rights of outsiders will have to be assessed against the preferential rights of the Norwegian coastal communities, Norwegian interests in reciprocal rights of access, and the responsible long term management of fish resources. In this context Norway is likely to negotiate a set of agreements based on reciprocal access rights or on gradual deescalation with the interested countries.[2] The process is almost as important as the substance in that it provides an early model for how the international community can accomplish the transition from a liberal to a managed regime of ocean management based on extensive coastal state rights and responsibilities.

The allocation of quotas will be a more difficult political issue: Inevitably any such system would put Norway and the Soviet Union in a preferential position due to the existing fishing effort and coastal state rights of possession.[3] It is important, however, that the quota agreements between Norway and the Soviet Union be based on reciprocal equity.[4] Unilateral Norwegian concessions to the Soviet Union could generate expectations of Soviet preferential rights in areas beyond Soviet jurisdiction which could have implications for other policy areas as well. The optics of this problem will be sharpened by the scaling down of the catches of traditional Norwegian allies among the European Community countries. However, reductions will apply to the DDR and Poland as well.

The problem is additive rather than specific. The political implications will be determined by the accumulated impact of the quota arrangements, the jurisdictional divisions, and the Svalbard regime. The danger is that of compartmentalized vision and management.

The Continental Shelf Division

This is not the place to review the legal issues involved in the Soviet–Norwegian dispute over the division of the continental shelf in the Barents Sea. Both countries are parties to the 1958 continental shelf convention. Norway contends that the division ought to follow the line of equidistance. The Soviet Union claims that special circumstances demand that the division be made in recognition of the sector 'principle'. The difference translates into some 155,000 km^2. The special circumstances to which the Russians make reference reportedly include their naval and strategic interests. The *premise* upon which any future agreement is based will be as important as the exact *line* that is drawn. If the premises be of such a nature and content that they logically apply also to the areas beyond the exact line of division, then their acceptance could amount to a concession of special interests in the area in general. The spectre of condominium would loom larger. Hence, for Norway the issue, in addition to its obvious 'territorial' components, is that of defining a clear boundary based on international law. Norway will presumably attempt to protect her rights to conduct exploitation of possible hydrocarbon resources in the shelf on her side of the demarcation line according to her own interests and in cognizance of international legal obligations (non-interference with maritime traffic, etc.). The time pressure on the negotiations derives from the establishment of EEZ's. Their establishment will necessitate a line of demarcation with respect to enforcement. 'Grey area' arrangements could have prejudicial effects on the continental shelf delimitation. In any event, Norway should strive to avoid temporary arrangements involving joint, as opposed to uniform, Soviet–Norwegian enforcement *vis-à-vis* their respective third party licencees. They should not enforce against each other's vessels. Any temporary arrangements may provide building blocks for 'permanent realities'.

Svalbard

The Svalbard issues are even more complex. The Norwegian legal position is clear enough. The major trouble is that it has failed to gain acceptance by some of the key signatories to the Svalbard Treaty, the Soviet Union, the United States, Britain and France. The issue is not primarily a legal one, although it is the legal arguments that form the political currency. Nobody really knows whether the continental shelf around Svalbard contains hydrocarbons which can be extracted on economic terms. That question is unlikely to be clarified for a long time. There is consequently no urgency in the issues involved. It is clear, however, that the Svalbard mining code is not a very appropriate instrument for governing possible off-shore activities around Svalbard. Hence, in the event of part of the shelf in the Barents Sea being included under the Svalbard Treaty it will be necessary to establish new rules for prospecting and production. The multilateral framework of the Svalbard regime could be emphasized by having such rules approved by a conference of the signatory powers.

It is the Norwegian contention that the continental shelf upon which Svalbard rests is a continuous geological extension of the Norwegian land mass which extends up to and beyond Svalbard. Hence, the issue has in fact been resolved by the definitions contained in the 1958 continental shelf convention and by the fact that Norwegian law is based solely on the exploitability criterion with respect to the delimitation of the continental shelf *vis-a-vis* the seabed. Furthermore, Norway holds that treaties cannot be interpreted extensively to delimit sovereign rights which have not been explicitly identified in the treaty. The Svalbard Treaty applies to the archipelago itself and its territorial sea of four nautical miles. The competing contention is that the rights of equal access to the natural resources of the archipelago already constitute such an extensive delimitation of Norwegian sovereignty that Norway cannot on the basis thereof acquire sovereign rights which are not similarly limited. The Svalbard Treaty establishes a nondiscriminatory regime and the spirit and intentions behind the treaty suggest that it should apply to the continental shelf as well. A possible compromise which would not jeopardize the Norwegian position with respect to the definition of the shelf or raise issues of delineation between metropolitan Norway and Svalbard would be to let the Svalbard rectangle, as defined in the treaty, determine the area to which the treaty applies offshore.

The more urgent issue is whether Norway can establish a 200 mile EEZ around Svalbard. It follows from the Norwegian legal position that it is the Norwegian view that such an act is within Norway's legal competence. However, it is not possible to argue that an EEZ has to be established for purposes of protecting the rights of an indigenous coastal population. Hence, the problem is that of *resource management* rahter than of *resource allocation*. That problem can become rather acute following the establishment of Norwegian and Soviet EEZ's, as such regimes would tend to transfer the fishing efforts of third countries to the Svalbard region where the young fish would be very vulnerable to depletion. The fishing effort around Svalbard has increased significantly over the past three years, primarily on the part of non-Norwegian vessels. Thus while the volume of the Norwegian catch exceeded that of all others combined in the period 1968–1973, it was less than half of the foreign catch in 1975. The Norwegian catch has remained fairly stable. Hence, it would seem advisable to contemplate a kind of nondiscriminating resource management regime around Svalbard with respect to fisheries. Such a solution would not prejudice positions with respect to the continental shelf.

It would seem important to obtain acceptance of the Norwegian position. That acceptance will not be based on the persuasive power of the Norwegian legal arguments. It will be based rather on Norway's ability to persuade the interested powers that it is in their own interests to have a regime managed equitably from Oslo rather than a competitive regime which could precipitate fears and moves which would stimulate great power rivalry in a sensitive area. The Norwegian Coast Guard which will be established[5] will constitute an important instrument which will have the function of making the Norwegian contentions and positions credible. But perceptions in this context will also be influenced by Norway's ability to avoid being drawn into a *de facto* Soviet-dominated condominium arrangement in the Barents Sea, and to ensure credible military reinforcements in an emergency. In this context it is useful to buy time, since credibility is a commodity which can only be accumulated by actual performance over time.

The Coast Guard Service may become important also for purposes of providing a credible exercise of Norwegian sovereignty and international obligations with respect to the Svalbard archipelago. The Svalbard Treaty does not contain provisions for general demilitarization. It does stipulate, however, that permanent naval installations are prohibited and that the area must not be used for warlike purposes. Norway has a responsibility for controlling adherence to the treaty, as well as respect for Norwegian territorial integrity in Svalbard. Moscow has protested against Norwegian naval visits to Svalbard and against Norway's landing military aircraft at the airport there. Such protests have been rejected. However, in future increased activity and attention on the part of the international community, as well as the need to patrol in a resource management zone around Svalbard, may make it desirable that elements of the Coast Guard Service at times operate from Svalbard. In this context a somewhat softer instrument than the regular Navy or Air Force may appear desirable for purposes of performing functions on and around Svalbard. Such considerations should partly determine the choice of aircraft and the fitting out of the vessels. Again the issues should be viewed in the broader context of establishing and maintaining an acceptable and credible regime of management in the high north of Europe.

The shaping of rules and activities in the north will be a very complex and demanding task. It will tax the foreign policy resources and stamina of a small country. The Soviet propensity for intransigence and heavy handedness in negotiations will be hard to handle.[6] In order to resist intimidation it will be necessary for Norway to obtain visible support from her primary allies.

Norway is not devoid of leverage. Both Norway and the Soviet Union need responsible management with respect to the major fish species in the north. Such management will require extensive Soviet–Norwegian cooperation. It is important that the management produce balanced and equitable results and that it be integrated into a larger regional management pattern for fisheries in the north-east Atlantic. Norway and the Soviet Union share, furthermore, an interest in preserving a state of low tension in a sensitive strategic area. Equitable and balanced cooperation, as opposed to Soviet-dominated exclusive bilateralism, can contribute to the general web of functional cooperation which, it is hoped, will enhance the protection of the process of East–West detente against reversal. Hence, the process of defining boundaries in the ocean areas of Europe's high north should be considered in the light of the principles laid down in the Final Act from the Conference on Security and Cooperation in Europe.

Notes

NEW STRATEGIC FACTORS IN THE NORTH ATLANTIC

1 For a more detailed discussion see 'New Naval Weapons Technologies' in *Strategic Survey 1975*, IISS, London 1976.
2 See Elizabeth Young, 'New Laws for Old Navies: Military Implications of the Law of the Sea', *Survival*, Nov./Dec. 1974.
3 For a fuller discussion see 'The European Military Balance' in *Strategic Survey 1975*, IISS, London 1976.

THE STATE AND FUTURE OF US NAVAL FORCES IN THE NORTH ATLANTIC

1 Figures for United States naval forces in this paper have been drawn from a number of public sources, notably the International Institute for Strategic Studies (London), the Center for Defense Information (Washington, DC), and *Jane's Fighting Ships 1975–76*. (London: Jane's Yearbook, 1975).
2 Quoted in Trygve Lie, *Hjemover* (Oslo: Tiden Norsk Forlag, 1958).
3 Richard L. Garwin, 'The Interaction of Antisubmarine Warfare with the Submarine-Based Deterrent', in Kosta Tsipis, Anne H. Gahn, Bernard T. Feld (eds.), *The Future of the Sea-Based Deterrent*, (Cambridge, Mass. and London: MIT Press, 1973), p. 89.
4 Kosta Tsipis, *Tactical and Strategic Anti-submarine Warfare*. (Cambridge, Mass. and London: MIT Press for Stockholm International Peace Research Institute, 1974), pp. 34–35.
5 For example, see Phillip A. Karber and Charles R. Wasaff, 'A Dissuasion Strategy for NATO', in Morton A. Kaplan (ed.), *NATO and Dissuasion*. (Chicago: University of Chicago Press, 1974).
6 James R. Schlesinger, *The Theater Nuclear Force Posture in Europe*. (U S Department of Defense, 1975), p. 18.
7 Rowland Evans and Robert Novak, 'Foreign Policy by Feud', *The Washington Post*, (November 1, 1975).
8 Martin Binkin and Jeffrey Record, *Where Does the Marine Corps Go from Here?* (Washington, D C: The Brookings Institution, 1976).
9 See notes 6–8.

THE STATE AND FUTURE OF THE SOVIET NAVY IN THE NORTH ATLANTIC

1 Office of the Chief of Naval Operations (OPNAV), *Understanding Soviet Naval Developments*, Washington, D.C.: Government Printing Office, 1975, p. 27.
2 Marshal of the Soviet Union A. A. Grechko, 'A Socialist, Multinational Army,' *Krasnaya Zvezda*, 17 December 1972.
3 Robert G. Weinland, 'Soviet Transits of the Turkish Straits – an Historical Note on the Establishment and Dimensions of the Soviet Naval Presence in the Mediterranean,' Arlington, Va.: Center for Naval Analyses, Professional Paper No. 94, April 1972 (Reprinted in: Michael MccGwire (ed.), *Soviet Naval Developments: Context and Capability*, New York: Praeger, 1973, pp. 324–343.)

4 Admiral Ephriam P. Holmes, USN, 'The Soviet Presence in the Atlantic,' *NATO Letter* 18-9 (September 1970), p. 11; Michael MccGwire, 'The Soviet Mediterranean Squadron, January 1968–June 1969: Deployment of Surface Combatants,' in MccGwire (ed.), *op. cit.,* pp. 382–386.

5 OPNAV, *op. cit.,* p. 14.

6 *Ibid.,* p. 13.

7 Robert G. Weinland, 'The Changing Mission Structure of the Soviet Navy,' Arlington, Va.: Center for Naval Analyses, Professional Paper No. 80, November 1971 (Reprinted in MccGwire (ed.), *op. cit.,* pp. 292–305.)

8 *U.S. Naval Institute Proceedings,* 96-11 (November 1970), p. 101.

9 George C. Wilson, 'Russians Keeping Missile-Subs Beyond U.S. Tracking Range,' *International Herald Tribune,* 29 April 1975; Phil Stanford, 'The Deadly Move to Sea, *New York Times Magazine,* 21 September 1975.

10 For an overview of Soviet naval activity see: Robert G. Weinland, 'Soviet Naval Operations – Ten Years of Change,' Arlington, Va.: Center for Naval Analyses, Professional Paper 125, August 1974 (Reprinted in MccGwire, Booth & McDonnell (eds.), *Soviet Naval Policy: Objectives and Constraints,* New York: Praeger, 1975, pp. 375–386.) For a recent discussion of Soviet Deployments to Cuba see: Barry M. Blechman and Stephanie E. Levinson, 'Soviet Submarine Visits to Cuba,' *U.S. Naval Institute Proceedings,* 101-9 (September 1975), pp. 30–39.

11 John W. Finney, 'Soviet Said to Use Guinea to Observe U.S. Shipping,' *New York Times,* December 6, 1973.

12 William Beecher, 'Soviet Missile Submarines on Patrol off U.S. Coasts,' *New York Times,* May 10, 1968; William Beecher, 'New Soviet Subs Relatively Noisy, Easy to Detect,' *New York Times,* October 9, 1969; Neil Sheehan, 'U.S. Sees Soviet Gain in Nuclear-Submarine Operations off the East Coast,' *New York Times,* October 4, 1970; Orr Kelly, 'Three Soviet Missile Subs Patrol off East Coast,' *Washington Star,* April 20, 1973.

13 William Beecher, 'U.S. and Soviet Craft Play Tag under Sea,' *New York Times,* May 11, 1968; Jack Anderson, 'U.S., Soviet Subs Scrape Hulls,' *Washington Post,* January 1, 1975.

14 For a desription of *Okean,* see: 'Soviet Maneuvers Summarized,' *U.S. Naval Institute Proceedings,* 96-11 (November 1970), p. 101 [translation of an article from the June 1970 edition of *La Revue Maritime*]. For a description of *Vesna,* see: 'Vast Soviet Naval Exercise Raises Urgent Questions for West,' *New York Times,* April 28, 1975; and 'Soviet Union: All the Ships at Sea,' *Time,* May 5, 1975, pp. 45, 46.

15 Lt. Charles L. Parnell, USN, " 'Sever' and the Baltic Bottleneck," *U.S. Naval Institute Proceedings,* 95-8 (August 1969), pp. 26–36.

16 Soviet exercise activity in the North Atlantic from 1960 through 1970 is described in some detail in: Admiral Ephriam P. Holmes, *op. cit.*

17 Admiral Ephriam P. Holmes, *op. cit.,* p. 10; Desmond Wettern, 'NATO Puzzled by Biggest Soviet Exercise,' *The Daily Telegraph* (London) 21 April 1975.

18 Charles G. Pritchard, 'The Soviet Marines,' *U.S. Naval Institute Proceedings,* 98-3 (March 1972), pp. 18–30.

19 Bradford Dismukes, 'Roles and Missions of Soviet Naval General Purpose Forces in Wartime: Pro-SSBN Operations?', Arlington, Virginia: Center for Naval Analyses, Professional Paper 130, August 1974 (Reprinted in: MccGwire, Booth & McDonnell, *op. cit.,* pp. 573–584.)

20 James M. McConnell, 'Military-Political Tasks of the Soviet Navy in War and Peace,' Arlington, Va.: Center for Naval Analyses, Professional Paper (forthcoming).

21 The rudiments of such a system might be in existence already. See, for instance:

Thomas O'Toole, 'Soviet Radar Satellites Tracking Surface Ships,' *Washington Post,* April 26, 1974; 'A New Soviet Submarine-Launched Missile,' *National Defense,* 54–236 (September–October 1974), p. 92; and Dave Polis, 'New Missile Threatens U.S. Fleet,' *San Diego Union,* October 6, 1975.

22 This is discussed at some length in: Admiral of the Fleet of the Soviet Union S. G. Gorshkov, 'Navies in War and Peace: Some Problems in Mastering the World Ocean,' *Morskoy Sbornik,* 2–73 (February 1973), pp. 13–25. For an Analysis of the series in which this article appeared, see: Robert G. Weinland, Robert W. Herrick, Michael MccGwire and James M. McConnell, 'Admiral Gorshkov's "Navies in War and Peace",' *Survival* 17-2 (March/April 1975), pp. 54–63.

23 Alvin Shuster, 'British Navy Ships Sent Near Iceland to Help Trawlers,' *New York Times,* May 20, 1973; 'Soviet Warships Shadow British Frigates off Iceland,' *Washington Post,* May 26, 1973.

CANADA AND NORTH ATLANTIC SECURITY

1 In 1970, the Ministry of Defence established a regional northern command centre in the Northwest Territories to improve the co-ordination of surveillance activities in the north.

2 The *Manhattan* was an American supertanker, specially equipped for sailing in Arctic waters, which was sent through the Northwest passage to prove the feasibility of sea transport of petroleum products in Arctic waters.

3 See, for example, Richard Rohmer, *Ultimatum: Oil or War?* (Markham, Ontario: Simon and Shuster, 1973).

4 Nils Ørvik, 'Defence Against Help – a Strategy for Small States?', *Survival,* September/October 1973, pp. 228–231.

5 Discussed earlier by this writer in *Alternativer for sikkerhet* (Oslo: E. G. Mortensen, 1970), pp. 161–172 and in Frans A. M. Alting von Geausau, *NATO & Security in the Seventies,* (Lexington: Heath, March 1971), pp. 73–102; also, 'Semi-Neutrality and Canadian Security', *International Journal,* Vol. XXIX No. 2 1972, pp. 186–215.

THE ARCTIC IN DANISH PERSPECTIVE

1 Finn Løkkegaard, *Det danske Gesandtskab i Washington 1940–52,* Copenhagen, 1968.

2 General Lucius D. Clay, 'North American Air Defense Command', *NATO's Fifteen Nations,* April–May 1974.

3 Cfr. SIPRI, *Tactical and Strategic Anti-submarine Warfare,* Stockholm, 1974.

4 Cfr. Palle Koch, *Grønland,* Copenhagen, 1975, p. 251.

5 Greenland Hearings, *Aarhus Stiftstidende,* Aarhus, 9 June 1968.

6 Chr. Vibe, *Arctic Animals in Relation to Climatic Fluctuations,* Copenhagen, 1967.

PROSPECTS FOR CONFLICT, MANAGEMENT, AND ARMS CONTROL IN THE NORTH ATLANTIC

1 For a comparative analysis of Soviet and American naval capabilities see Johan Jørgen Holst, 'The Navies of the Super-Powers: Motives, Forces, Prospects', in *Power at Sea, Part II: Super-powers and Navies,* Adelphi Paper No. 123 (London: International Institute for Strategic Studies, 1976).

2 See Johan Jørgen Holst, *Den amerikanske garantien i norsk sikkerhetspolitikk* (Oslo: Den norske Atlanterhavskomité, 1975).

3 For an analysis of an operation which might have produced involuntary confrontation see Johan Jørgen Holst, 'Signals, Decisions and Strategy: The Submarine Hunt

in the Sognefjord 1972', *Cooperation and Conflict*, IX (4), 1974, pp. 297–311.

4 For an examination of Norwegian requirements see *Oppsynet med fiskeri- og petroleumsvirksomheten*, NOU 1975:50 (Forsvarsdepartementet).

5 The Norwegian policy rationale is outlined in *St.meld. nr. 25 (1973–74) Petroleumsvirksomhetens plass i det norske samfunn* (Finansdepartementet) and *St.meld. nr. 30 (1973–74) Virksomheten på den norske kontinentalsokkel m. v.* (Industridepartementet). Both parliamentary documents are available in English translation.

6 For elaboration see Johan Jørgen Holst, *Oljen i sikkerhetspolitikken* (Oslo: Den norske Atlanterhavskomité, 1975).

7 *Aftenposten*, 11 September 1975.

8 For an assessment of the global issues of ocean policy see Douglas M. Johnston, Michael Hardy, Ann L. Hollick, Johan Jørgen Holst, Shigeru Oda and Richard N. Cooper, *A New Regime for the Oceans* (New York: The Trilateral Commission, 1975).

9 See Johan Jørgen Holst, 'The Strategic and Security Requirements of North Sea Oil and Gas' in Martin Sæter and Ian Smart (eds.), *The Political Implications of North Sea Oil and Gas*, (London: IPC Science and Technology Press, 1975), pp. 131–144.

10 *Sikkerhetsforskrifter for petroleumsproduksjon på kontinentalsokkelen*, NOU 1975: 43 (Industridepartementet).

11 See, e.g., Barry M. Blechman, *The Control of Naval Armaments*, (Washington DC: The Brookings Institution, 1975).

12 See Johan Jørgen Holst (ed.), *Five Roads to Nordic Security* (Oslo: Universitetsforlaget, 1973).

THE STRATEGIC IMPORTANCE OF THE NORTH ATLANTIC: NORWAY's ROLE AND OPTIONS

(Notes added January 1977)

1 Such a zone was established on January 1, 1977. Cfr. *Ot.prp. nr. 4 (1976–77) Om lov om Norges økonomiske sone*, and *Kgl. res. av 17.12.1976 Om iverksettelse av Norges økonomiske sone*.

2 Norway concluded fishing agreements with the following countries in 1976: DDR, Finland, Iceland, Poland, Sweden and the USSR. Negotiations were under way with the European Community as well as with Portugal and Spain.

3 The Soviet Union announced the establishment of a 200 mile EEZ on December 10, 1976.

4 A Soviet–Norwegian protocol concerning TAC (Total Allowable Catch) and quotas in their 200 mile EEZs was concluded December 24, 1976. The protocol covers arctic cod, haddock, halibut, redfish, seithe, and capelin (no TAC). The distribution of the Soviet and Norwegian quotas with respect to the zones as well as the distribution of the third country quota among the countries involved remained. 'Temporary Regulations with Respect to Foreign Fisheries in the Norwegian Economic Zone' were decreed by the Norwegian Department of Fisheries December 27, 1976, and temporary rules of enforcement by the Ministry of Defence December 28, 1976.

5 *St.meld. nr. 81 (1975–76) Oppsynet med fiskeri- og petroleumsvirksomheten: Etablering av en kystvakt*.

6 The Soviet Union carried out four sets of tests involving the firing of (obsolescent) ICBMs (SS-7's?) into the disputed area in the Barents Sea during 1976. It seems clear that Moscow could have chosen a less controversial impact area for fulfilling her SALT-I obligations.

Notes on Contributors

CHRISTOPH BERTRAM (b. 1937) is the Director of the International Institute for Strategic Studies in London. A German citizen, he was educated in Berlin, Bonn (Dr of Law 1967) and Paris. He joined the Institute in 1967 as a Research Associate, and its directing staff in 1969. After a brief period with the Planning Staff of the German Ministry of Defence, he rejoined the Institute and has been its Director since 1971.

BJØRN BJARNASON (b. 1944) obtained his cand. jur. (Law) degree from the University of Iceland in 1971. He has been the President of the Icelandic Union of Students and an editor of the book-publishing house Almenna Bokfelagid. Mr Bjarnason has been Foreign News Editor of the Reykjavik daily *Visir*. Since October 1974 he has been employed in the office of the Prime Minister where he now occupies the position of Deputy Secretary General. He has written extensively on Icelandic foreign policy.

ERLING BJØL (b. 1918) is Professor of International Relations at the Institute of Political Science, Aarhus University. He is the author of many books and articles on a variety of foreign policy questions. Among his works are counted books on the strategy of development in the Italian Mezzagiorno, French policy towards Europe, a general introduction to the study of international relations, a three-volume history of the post-war world, as well as several studies of Danish foreign and defence policy.

JOHAN JØRGEN HOLST (b. 1937), a political scientist, was educated at Columbia University and the University of Oslo. He has been a Research Associate at the Norwegian Defence Research Establishment, the Center for International Affairs of Harvard University and at Hudson Institute. Since 1969 he has been Director of Research at the

Norwegian Institute of International Affairs. In January 1976 Mr Holst was appointed Under-Secretary of State for Defence in Norway. He has published and lectured widely on subjects relating to international relations, strategy and arms control. Mr Holst is a member of the council of the International Institute for Strategic Studies.

PHILLIP A. KARBER (b. 1946) is Vice President for National Security Programs and Director of Strategic Studies at the BDM Corporation in Vienna, Virginia. In this capacity he has headed research in the military policy and behavioural science areas. He is a contributing author to several recent books on foreign and defence policy issues.

JON L. LELLENBERG (b. 1946) is a Senior Political Analyst, Strategic Forces and Assessment with the BDM Corporation in Vienna, Virginia, specializing in European military and political issues. Mr Lellenberg has written extensively on developments in the North Atlantic.

GEORGE LINDSEY (b. 1920) is Chief of the Operational Research and Analysis Establishment in the Department of National Defence, Ottawa. He has degrees from Toronto, Queens, and Cambridge, and has worked in the areas of nuclear physics, operational research, and strategic studies. He is the co-author of a book on the nuclear balance.

D. P. O'CONNELL, R.D., Ph.D., LL.D., D.C.L., F.R. Hist.I, (b. 1924) is Chichelle Professor of Public International Law in the University of Oxford, Fellow of All Souls College, Associé de l'Institut de Droit International and Commander, Royal Naval Reserve. Professor O'Connell was born in New Zealand and received his education in New Zealand and at Trinity College, Cambridge. He was formerly Professor of International Law in the University of Adelaide, Australia, consultant on the Law of the Sea to several governments, and a delegate at the Law of the Sea Conference.

ANDERS C. SJAASTAD (b. 1942) is Director of Information at the Norwegian Institute of International Affairs (NIIA), Oslo. He has been a Research Assistant at the University of Oslo and Research Associate at the NIIA. Mr Sjaastad is the author of several articles on foreign policy issues as well as co-author of books on departmental decision-making in Norway and on security and politics in the Norwegian Sea area.

192

JOHN KRISTEN SKOGAN (b. 1942) is a Research Associate at the Norwegian Institute of International Affairs. He has been a Research Assistant at the Norwegian Defence Research Establishment and a research associate at the International Institute for Strategic Studies, London. Mr Skogan has authored several studies in the area of foreign policy analysis. He is the co-author of a book on security and politics in the Norwegian Sea area.

FINN SOLLIE (b. 1928) is the Director of the Fridtjof Nansen Foundation at Polhøgda, Oslo, where he has been employed since 1965. Previously he was a member of the Norwegian Foreign Service, serving in Oslo and Washington. His recent work has been concentrated on international problems related to the exploration and exploitation of 'new territories', i.e. sea and seabed areas and polar regions. He has written extensively in books and periodicals.

ROBERT G. WEINLAND is a Senior Fellow and member of the Defense Analysis Staff at the Brookings Institution. Before joining Brookings in 1975, he was a member of the professional staff of the Center for Naval Analyses (CNA). During nearly ten years at CNA he participated in and directed several studies of the Soviet Navy and naval operations. He also directed a study of Western Alliance processes. During 1974–75 he served in London as CNA Representative on the staff of the Commander-in-Chief, US Naval Forces, Europe. He has published a number of articles on the Soviet Navy.

NILS ØRVIK (b. 1918) is Director of the Centre of International Relations and Professor of Political Studies at Queens University, Kingston, Ontario. Born in Norway, Dr Ørvik taught at the University of Oslo before emigrating to Canada. He has held fellowships at the University of Wisconsin, London School of Economics and Political Science, Harvard University, University of California and Columbia University. Dr Ørvik is the author of several books on Norwegian foreign and defence policies and he has published and lectured widely in the field of foreign policy analysis.